Salmonella - Perspectives for Low-Cost Prevention, Control and Treatment

Edited by Hongsheng Huang and Sohail Naushad

Published in London, United Kingdom

Salmonella - Perspectives for Low-Cost Prevention, Control and Treatment
http://dx.doi.org/10.5772/intechopen.102150
Edited by Hongsheng Huang and Sohail Naushad

Contributors
Cornelius Cano Ssemakalu, Manuela Oliveira, João Bettencourt Cota, Madalena Vieira-Pinto, Leonard I. Uzairue, Olufunke B. Shittu, Ruimin Gao, Jasmine Rae Frost, Sohail Naushad, Hongsheng Huang, Dele Ogunremi, Joseph K.N. Kuria, Hongxia Zhao

First published in London, United Kingdom, 2024 by IntechOpen
IntechOpen is the global imprint of INTECHOPEN LIMITED, registered in England and Wales, registration number: 11086078, 5 Princes Gate Court, London, SW7 2QJ, United Kingdom

British Library Cataloguing-in-Publication Data
A catalogue record for this book is available from the British Library

Additional hard and PDF copies can be obtained from orders@intechopen.com

Salmonella - Perspectives for Low-Cost Prevention, Control and Treatment
Edited by Hongsheng Huang and Sohail Naushad
p. cm.
Print ISBN 978-1-83962-473-5
Online ISBN 978-1-83962-474-2
eBook (PDF) ISBN 978-1-83962-500-8

For EU product safety concerns:
IN TECH d.o.o., Prolaz Marije Krucifikse Kozulić 3, 51000 Rijeka, Croatia,
info@intechopen.com or visit our website at intechopen.com.

Meet the editors

Hongsheng Huang obtained a BVM from the College of Veterinary Medicine, Inner Mongolia Agriculture University, People's Republic of China; an MPhil in Veterinary Microbiology and Immunology from the University of Edinburgh, Scotland; and a Ph.D. in Veterinary Immunology, Microbiology and Pathology from the University of Saskatchewan, Canada. He is currently a research scientist at Ottawa Laboratory – Fallowfield, Canadian Food Inspection Agency. His research interests include the development of methods to isolate, detect, and characterize foodborne bacterial pathogens using various approaches, including conventional bacterial culture methods, immunological and molecular methods, nanoparticle biosensors, and genomics tools; to detect and characterize antibiotic-resistant genes in food and environment; and to detect and inactivate abnormal prion protein in animals.

Dr. Sohail Naushad obtained a Ph.D. in Biochemistry from McMaster University, Canada, in 2015, where he identified novel molecular markers for various pathogens, such as foodborne pathogens, *Enterobacteriales, Pasteurellales, and Xanthomonadales*, which affect humans, animals, and plants. After completing his Ph.D., he served as a postdoctoral fellow at the University of Calgary, Canada, from 2015 to 2018, during which he was supported by the Eyes High postdoctoral fellowship and the NSERC CREATE in the Milk Quality postdoctoral award. During his fellowship, he discovered a new bacterial species, *Staphylococcus debuckii*, and developed algorithms, statistical methods, datasets, and pipelines for analyzing whole-genome sequencing data to understand the phylogeny, virulence, and antimicrobial resistance of bovine pathogens. Dr. Naushad is currently working as a research scientist at the Canadian Food Inspection Agency, Ontario, Canada, where he conducts metagenomic and microbiological research. He is involved in the development of a mobile device for the field detection of microbial threats in Canadian food chains and environments using sequencing platforms such as MinION, MiSeq, and Ion Torrent Ion S5.

Contents

Preface

Salmonella, a Gram-negative bacterium belonging to the *Enterobacteriaceae* family, can cause infections in humans and animals, and is one of the most common causes of bacterial gastroenteritis. According to the World Health Organization, more than 2 billion people worldwide suffer from diarrheal diseases annually and one of four of these diseases is caused by *Salmonella* each year.

Since the discovery of *Salmonella* in the late 1800s, great progress has been made in understanding its genetics, classification, pathogenesis, detection, prevention, control, and treatments. Numerous reviews and book chapters on *Salmonella* have been published. However, some gaps remain to be addressed. This book presents seven chapters that focus on low-cost prevention, control, and treatment of the salmonellosis in developing countries.

The nomenclature of *Salmonella* is complicated and constantly evolving. Chapter 1 provides an updated review of *Salmonella* nomenclature. Currently, it is commonly recognized that the genus *Salmonella* contains two species: *S. enterica* and *S. bongori* based on the genomic relatedness, with *S. enterica* containing six subspecies (subsp.), *enterica* (I), *salamae* (II), *arizonae* (IIIa), *diarizonae* (IIIb), *houtenae* (IV), *and indica* (VI). *Salmonella* isolates are further serotyped using the Kauffmann–White scheme with more than 2600 serovars identified. In addition, three pathogenic serotypes, *Paratyphi C*, Dublin, and Typhi, are identified based on a special subtype of heat-sensitive K antigen at the bacterial capsular surface.

There are two types of human salmonellosis: (1) gastroenteritis (non-typhoid salmonellosis [NTS]), a localized infection due to ingestion of contaminated food or water, and (2) enteric fever (typhoid fever), a severe and life-threatening systemic infection. Only a small fraction of serovars is associated with human infections, mostly belonging to *S. enterica* subspecies *enterica*, which is composed of more than 1500 serovars. Two of these serovars, *S. typhimurium* and *S. enteritidis*, are responsible for more than 99% of NTS. Typhoid fever is caused mainly by S. typhi, and a clinically indistinguishable condition caused by S. Paratyphi A. Typhoid fever remains a global health problem, especially in the developing countries with substandard water supplies and poor sanitation. Certain *Salmonella* such as *Salmonella enterica* subsp. *salamae*, subsp. *arizonae*, subsp. *diarizonae, houtenae*, *S. enterica* subsp. *indica*, and *S. bongori* are usually isolated from cold-blooded animals and the environment with a potential to cause disease in humans.

Animals are major reservoirs of NTS. As reviewed in Chapters 1, 2, and 3, the transmission of NTS infection to humans can occur through the ingestion of food or water contaminated with infected animal waste, direct contact with infected domestic, wild, and companion animals, or consumption of infected animal food products, and direct contact with the contaminated environment. S. typhi causing typhoid fever can be transmitted through contaminated food and water or through close contact with an infected person. The diversity of possible reservoirs of infection leads to significant challenges for public health authorities in controlling infections. Chapters 5 and 6 review low-cost measures to prevent and control *Salmonella* in water and animals using either solar disinfection for water or biocide for animals, which could be practical for low-income countries.

Most people recover from salmonellosis without any specific treatment. Antibiotics are typically used only to treat patients with severe illness caused by *Salmonella* infections. Since the report of the first incidence of *Salmonella* resistance to chloramphenicol in 1960s, the emergence of antimicrobial resistance towards single or multi drugs in *Salmonella* strains has become a serious health problem worldwide, particularly in Africa and Asia, as reviewed in Chapters 3 and 7. Interestingly, Chapter 7 also compares AMR in *Salmonella* with other microbes and introduces a new concept of AMR reversal using traditional Chinese medicine as alternatives for treatment, which could lead to new strategies for clinical treatment of bacterial infections.

Ensuring the safety of water and food, predominantly poultry, eggs, and dairy products, is the main strategy for eliminating possible transmission routes of typhoid *Salmonella* as well as NTS. Detecting and characterizing the isolates is crucial for epidemiology and prevention of salmonellosis. Chapter 4 reviews the research progress for understanding the roles played by CRISPR-Cas systems in *Salmonella* immune response, as well as genome editing and its potential for pathogen typing in diagnosis and surveillance.

We believe that the information provided in this book will encourage *Salmonella* researchers, medical professionals, and students to further enhance their own research and education, and encourage new researchers to include *Salmonella* in their future research initiatives. We are grateful to various researchers and scientists across the world who have contributed to this book and hope that the information provided will be well received in the *Salmonella* field, particularly in developing countries, and beyond.

Hongsheng Huang and Sohail Naushad
Canadian Food Inspection Agency,
Ottawa Laboratory – Fallowfield,
Ottawa, Ontario, Canada

Chapter 1

Salmonella: A Brief Review

Sohail Naushad, Dele Ogunremi and Hongsheng Huang

Abstract

Salmonella causes significant illness in humans and animals and is a major public health concern worldwide, contributing to an increased economic burden. *Salmonella* is usually transmitted through the consumption of contaminated food, such as raw or undercooked meat, poultry, eggs, and dairy products, and water or through contact with infected animals or their environment. The most common symptoms of salmonellosis, the illness caused by *Salmonella*, include diarrhea, fever, and abdominal cramps; in severe cases, the infection can lead to hospitalization and even death. The classification and taxonomy of *Salmonella* were historically controversial, but the genus is now widely accepted as composed of two species and over 2600 serovars. Some of these serovars infect a single host, that is, host-restricted, whereas others have a broad host range. Colonization of the host is complex and involves a series of interactions between the *Salmonella* and the host's immune system. *Salmonella* utilizes an array of over 300 virulence factors, mostly present in *Salmonella* pathogenicity islands (SPIs) to achieve adherence, invasion, immune evasion, and, occasionally, systemic infection. Once colonized, it secretes a number of toxins and inflammatory mediators that cause diarrhea and other symptoms of salmonellosis. The overuse and misuse of antibiotics in human and animal medicine and agriculture have contributed to the development of antimicrobial resistance (AMR) in *Salmonella*, making AMR strains more severe and difficult to treat and increasing the risk of morbidity and mortality. Various methods are used for the detection of *Salmonella*, including traditional culture methods, molecular methods such as polymerase chain reaction (PCR) and loop-mediated isothermal amplification (LAMP), and immunological-based assays. Because of its ubiquitous distribution, the prevention and control of *Salmonella* transmission remain a significant challenge. This chapter briefly covers the history, classification, transmission, pathogenesis and virulence factors, antimicrobial resistance genes, detection, diagnosis, surveillance, prevention, and control pertaining to *Salmonella*.

Keywords: *Salmonella*, history, taxonomy, classification, transmission, pathogenesis, virulence factors, antimicrobial resistance genes, *Salmonella* detection, diagnosis, surveillance, prevention and control

1. Introduction

Salmonella is a bacterial genus consisting of many closely related organisms, which remains a major cause of morbidity and mortality worldwide with significant public health implications, contributing to the economic burdens of both developed and economically marginalized countries because of costs associated with monitoring, surveillance, prevention, and treatment of salmonellosis [1–7]. Karl Joseph Eberth of the University of Zurich, a physician and pathologist, described a bacillus in the abdominal lymph nodes and spleen of a patient who died of typhoid in 1879 [8]. At the time, the bacterium was referred to as Eberth's *Bacillus* [9, 10] and followed by the discovery of *Bacillus* as the cause of human typhoid fever by George Gaffky in 1884 [1, 8, 10]. Nevertheless, the genus "*Salmonella*" was named after Daniel Elmer Salmon, an American veterinary pathologist and head of the United States Department of Agriculture (USDA) Microorganism Research Program in the late 1800s [1]. Together with Theobald Smith, Salmon isolated *Salmonella* from the intestines of the pigs that succumbed to the disease known as hog cholera in 1884 [8, 11]. Historians and scientists studying past disease outbreaks have concluded that many catastrophic disease outbreaks of the early ages were likely caused by *Salmonella*, more specifically, typhoid infections [5, 12]. As early as 430 B.C., a plague, which is now believed to have been typhoid fever, wiped out a third of the population of Athens [1, 5].

Salmonellosis is a common cause of foodborne illness in the world [2, 3, 13]. Symptoms of salmonellosis can include fever, diarrhea, abdominal cramps, and vomiting and can last for several days. According to the World Health Organization (WHO), more than 2 billion people worldwide suffer from diarrheal diseases annually [8], and 1 of 4 of these diseases is caused by *Salmonella* [4, 7, 8, 14]. Depending on host factors and the serotype of *Salmonella,* as well as the presence of antimicrobial resistance (AMR) genes, 11–20 million cases of salmonellosis become severe and life-threatening, leading to 161,000 deaths annually [4, 7]. According to the Centers for Disease Control and Prevention (CDC), salmonellosis is one of the most common bacterial foodborne illnesses in the United States, with an estimated 1.35 million cases occurring annually [15]. The Public Health Agency of Canada (PHAC) estimates that there are about 87,500 cases of salmonellosis each year in Canada [16]. According to the European Centre for Disease Prevention and Control (ECDC) in 2021, salmonellosis was the second most common bacterial foodborne infection in Europe, with an estimated 60,050 cases occurring annually [15]. In developing countries, the burden of salmonellosis is usually higher because of the combination of many factors, such as poor hygiene and sanitation conditions, lack of access to safe water and proper food handling practices, lack of proper disease reporting structure, and limited resources for disease surveillance and response [4, 14, 17].

This chapter will briefly review the nomenclature, transmission, pathogenesis, diagnosis and detection, prevention, control, and treatment.

2. The organism

Salmonella is a rod-shaped, Gram-negative bacterium and consists of a cell wall, cell membrane, cytoplasm, ribosomes, plasmids, and nucleoid region. It has a diameter of around 0.7 to 1.5 μm, a length from 2 to 5 μm, and flagella, which allows for motility [18]. *Salmonella* is a chemoorganotroph, which means it obtains energy from the oxidation of the reduced organic compounds, and is a facultative anaerobe [18]. After colonizing the epithelium, *Salmonella* reproduces by binary fission, which begins with the replication and attachment of the DNA molecules to the cell membrane. Once the bacterium doubles its original size, the cell membrane begins to pinch inward and a cell wall forms between the two DNA molecules to divide the original cell into two identical daughter cells [18, 19]. Once reproduced, the bacterium either stays within the intestine or enters the bloodstream or lymph tracts [19]. *Salmonella* can also survive for several weeks outside of a living host in a dry environment and several months in water and is often not destroyed by freezing temperatures. The bacteria will only be destroyed in temperatures above 75°C, which makes raw and undercooked food, together with improperly washed fruits and vegetables, a common source of transmission of the bacterium.

The genome of *Salmonella* is relatively small, with a size ranging from 4.7 to 5.3 million base pairs [20–22]. It is a single circular chromosome that encodes a wide range of proteins involved in various cellular processes, including metabolism, regulation, and pathogenicity [22–25]. Several studies have characterized the genomic features of *Salmonella* and identified genes that are important for its survival, virulence, and antimicrobial resistance [23, 25–29]. These include genes encoding toxins and other virulence factors that allow the bacteria to colonize and infect host tissues, as well as genes involved in the uptake and metabolism of nutrients. One of *Salmonella's* most well-known genomic features is the presence of prophages, which are bacteriophages (viruses that infect bacteria) that have integrated into the bacterial genome [30, 31]. Prophages can be activated under certain conditions, leading to the production of new phages that can potentially spread to other bacteria. Genomic studies of *Salmonella* have also identified a number of genes that are involved in antibiotic resistance [32–35]. These genes can be horizontally transferred among bacteria, leading to the spread of antibiotic resistance among different bacterial species. In recent years, the use of whole-genome sequencing (WGS) to study *Salmonella* has allowed for a deeper understanding of the organism's epidemiology, biology, evolution, and population structure [36–39]. WGS can accurately predict various characteristics and traits of a *Salmonella* isolate based on its genomic sequence, replacing the need for time-consuming and costly traditional methods [26, 30, 39].

3. Classification

Salmonella belongs to the Kingdom *Monera* or *Eubacteria,* phylum *Proteobacteria,* class *Gamma-Proteobacteria,* order *Enterobacteriales*, family *Enterobacteriaceae*, and the genus *Salmonella* [40, 41]. The nomenclature of the genus *Salmonella* has been confusing and controversial and has two systems of nomenclature widely used for the taxonomical assignments of *Salmonella*. One system, which does not conform to the rules of the Bacteriological Code but has wide acceptance, was proposed in the 1980s by Le Minor and Popoff in 1980 [18, 42], whereas the second system conforms to the rules of the Bacteriological Code and is not widely used [43]. To resolve the discrepancies in the taxonomical system of *Salmonella*, the Judicial Commission of the International Committee on the Systematics of Prokaryotes issued an Opinion (Opinion 80), with the intention that it should solve these discrepancies [41]. However, Opinion 80 was also limited to matters of nomenclature and meant to provide a clear presentation and interpretation of *Salmonella* taxonomy of the widely accepted division of the genus *Salmonella* into two species [41]. There are approximately 2000 similar species of *Salmonella*, which has caused much confusion in terms of classifying each species. In order to simplify this, the Center for Disease Control and Prevention (CDC) has agreed upon two species of *Salmonella*, *Salmonella enterica* and *Salmonella bongori*. *Salmonella enterica* is subdivided into six subspecies: *S. enterica* ssp. *enterica* (I), *S. enterica* ssp. *salamae* (II), *S. enterica* ssp. *arizonae* (IIIa), *S.*

enterica ssp. *diarizonae* (IIIb), *S. enterica* ssp. *houtenae* (IV), and *S. enterica* ssp. *Indica* [44, 45]. These species and subspecies are further classified into multiple serotypes based on the White–Kauffmann–Le Minor scheme updated by the World Health Organization's Collaborating Centre for Reference and Research on *Salmonella* at the Pasteur Institute, Paris, France [45, 46]. The genus *Salmonella* is made up of approximately 2600 serovars based on antigenic polymorphisms of their somatic O antigens (lipopolysaccharide), H antigens (flagellar proteins), and Vi antigens (capsular polysaccharides; [41, 43]). Most of the serovars belong to *S. enterica* ssp. *enterica* (I), and the most common serogroups are A, B, C1, C2, D, and E [43].

To avoid confusion in writing names and differentiate between serovars designation and species-level designation, it is recommended to write them in Roman style starting with a capital letter [43]. The current convention used in scientific writing is to state first the genus name and then the species name, followed by the word "serovar" (which can be abbreviated as "ser."), and finally the actual name of the serovar [43]. An example, at first, is to write *Salmonella. enterica* subsp. *enterica* serovar (or ser.) Typhimurium. To simplify this long written convention and avoid the long nature of nomenclature, the name can be shortened by writing the genus name, followed directly by the serovar name starting with a capital letter; for example, *Salmonella. enterica* subsp. *enterica* serovar Typhimurium can be written as *Salmonella* Typhimurium.

Salmonella has a wide host range, which based on host adaptability, can be divided into three broad groups [47]. Group 1 *Salmonella* serovars are adapted to humans and higher primates such as *Salmonella* Typhi, *Salmonella* Paratyphi A, B, C, and *Salmonella* Sendai [47]. Group 2 *Salmonella* are largely adapted to specific animal hosts such as *Salmonella* Dublin in cattle, *Salmonella* Gallinarum in poultry, *Salmonella* Abortusequi in horses, *Salmonella* Abortusovis in sheep, and *Salmonella* Choleraesuis in pigs [47]. Group 3 *Salmonella* have a wide host range including humans, animals, and the environment, such as *Salmonella* Typhimurium and *Salmonella* Enteritidis, the two most common serotypes of *Salmonella* transmitted to humans in most parts of the world [47].

4. Transmission

Salmonella generally resides in the gut of animals, including birds, and is usually transmitted to humans by eating contaminated foods [7, 15]. These contaminated foods are typically from animal origin such as beef, poultry, milk, or eggs, but all food types including vegetables may be contaminated [15, 16]. The bacteria are commonly found in raw eggs and undercooked chicken and eggs. Person-to-person spread is possible in close contact, especially during the acute diarrheal phase of the illness [15, 16, 48]. *Salmonella* is transmitted by the consumption of raw food that is contaminated with the bacteria, such as vegetables that have not been cooked or washed properly, meat, or eggs. *Salmonella* can be transferred if the food handler or processor does not use gloves when dealing with food [15, 16]. It can also be transmitted by reptiles or rodents through their feces [49]. If the food is contaminated with a high concentration of *Salmonella*, the person is more likely to become infected. Children, elderly people, and HIV-positive people are more likely to become infected [7, 16, 49]. Once ingested, *Salmonella* embeds itself into the intestinal epithelium where it reproduces [50]. The liver, spleen, and especially the gall bladder have a high concentration of *Salmonella*.

If left untreated, the organism can travel through the bloodstream to joints, organs, placenta, and membranes around the brain [50]. The toxins released by the bacteria can damage various organs in the body [51, 52].

5. Diseases

The general term for infections caused by *Salmonella* is salmonellosis, which is generally is divided into two main types: typhoidal and non-typhoidal. Typhoidal salmonellosis or typhoid fever is caused mainly by *Salmonella* Typhi, characterized by symptoms such as fever, weakness, abdominal pain, and loss of appetite [1, 4, 53] and typically acquired through the consumption of contaminated food or water and more common in developing countries [4]. However, non-typhoidal salmonellosis is caused by a variety of *Salmonella* serotypes and mostly causes food poisoning symptoms, such as diarrhea, abdominal cramps, and fever and is more common in developed countries [14, 53, 54]. Most of the people infected with *Salmonella* will develop diarrhea, abdominal cramps, fever, and vomiting, which can last up to a week [1, 4, 55, 56]. Other symptoms caused by *Salmonella* infection include the enlargement of the spleen and lymph nodes, accumulation of fluid and blood in organs such as the lungs, and damage to the liver [1, 53, 55]. In chronic cases, arthritis may even occur, known as Reiter's Syndrome, and can last for months or even years [57, 58]. Different symptoms will occur in different mammals and birds.

6. Pathogenesis and virulence factors

Salmonella uses an array of virulence genes as part of its mechanism of pathogenesis [59]. These genes encode proteins that help the bacteria to evade the host's immune system, colonize and survive in host tissues, and cause inflammation and tissue damage [55, 60, 61]. Understanding the virulence genes of *Salmonella* can help researchers to develop strategies for preventing and treating infections caused by these bacteria. Some key virulence genes involved in each step of *Salmonella* pathogenesis include the following:

- *invA*: encodes a protein called Invasin, which is involved in the invasion of host cells by *Salmonella* [60, 62].

- *spvC*: encodes a protein called SpvC, which is involved in the formation of a specialized structure called the *Salmonella*-containing vacuole (SCV) within host cells. The SCV helps *Salmonella* to evade the host's immune system and establish an intracellular infection [63].

- *sopE*: encodes a protein called SopE, which is involved in the manipulation of host cell signaling pathways. SopE can activate signaling pathways that promote inflammation and tissue damage, as well as inhibit signaling pathways that would otherwise inhibit bacterial growth [64].

- *sseL*: encodes a protein called SseL, which is involved in the secretion of toxins into host cells, macrophage killing, and enhancement of virulence [65–67].

Many of these virulence genes are located on pathogenicity islands known as *Salmonella* pathogenicity islands (SPIs), which are thought to be acquired by horizontal gene transfer [68, 69]. SPIs are regions of bacterial DNA found in some strains of *Salmonella* and believed to play a role in the bacteria's ability to cause diseases [68, 69]. SPIs are typically composed of several genes, including virulence genes that encode proteins involved in the bacterium's ability to invade host cells, evade the immune system, and survive in different environments [69]. There are a total of 24 SPIs (1–24) recognized in *Salmonella* so far [70]. Each SPI is believed to have a specific function in the pathogenesis of *Salmonella* infections [68, 69]. For example, SPI-1 is involved in the bacterium's ability to invade and replicate within host cells [70], whereas SPI-2 is involved in the production of a toxin that can cause inflammation in the intestinal tract [69]. SPI-1 is a large and complex region of DNA, comprising approximately 40 genes [69]. Many of these genes are involved in the production of proteins called effectors, which are secreted by the bacteria into host cells and function to alter host cell function [70]. For example, some effectors can disrupt the normal functioning of the host cell's cytoskeleton, enabling the bacteria to move within the host tissue and evade immune cells [69, 70]. Other effectors can interfere with the host cell's signaling pathways, helping the bacteria to evade detection by the host's immune system. The type III secretion system (T3SS) encoded by SPI-1 is considered to be the most important virulence factor for *Salmonella* [68–70]. SPI-2 is another 40 kb long region of DNA found in certain strains of *Salmonella* bacteria, which has two distinct regions encoding proteins required to establish and maintain *Salmonella*-containing vacuole essential for *Salmonella* replication [71]. SPI-2 encodes a second T3SS, implicated in systemic pathogenesis [72]. The two regions of SPI-2 have unique species-specific distribution; for example, the larger 25 kb region is exclusive to *S. enterica*, whereas a second 15 kb long region is identified in *S. bongori* [69]. Like SPI-1, it contains a number of genes that contribute to the pathogenicity of the bacteria; SPI-2 is a smaller and less complex region of DNA than SPI-1 [69].

SPI-3 is a 17 kb long chromosomal DNA region that encodes many proteins involved in adhesion, such as MisL protein, which is vital for the long-term persistence of *Salmonella* [69]. SPI-3 is thought to be conserved in *S.* Typhi and *S.* Typhimurium [69, 73]. Similarly, SPIs 4–24 are involved in various aspects of pathogenesis and functions, all of which are not fully understood yet [69]. However, understanding the role of SPIs in *Salmonella* pathogenesis is important for the development of vaccines and therapies against the bacteria. Researchers are currently studying the mechanisms by which *Salmonella* utilizes SPIs to cause diseases, with the goal of finding new ways to prevent or treat infections caused by this bacterium.

Various *Salmonella* strains also contain plasmids, which have virulence and AMR genes [74–77]. *Salmonella* plasmids are usually small, circular pieces of DNA that are found in some strains of *Salmonella*. However, some strains also carry large *Salmonella* virulence plasmids [22, 77, 78]. Plasmids are separate from the bacterial chromosome and can carry a variety of genes, including those that confer antibiotic resistance or other traits that can help the bacteria survive and thrive in different environments [22, 79–81]. Some *Salmonella* plasmids carry virulence genes, which are responsible for the bacteria's ability to cause illness in humans and animals. *Salmonella* plasmids can be transmitted from one bacterium to another through horizontal gene transfer and can contribute to the evolution of new pathogenic strains [74, 82, 83]. Plasmids are an important tool in molecular biology and are often used to introduce new genes into bacterial cells for research or biotechnology purposes. There are several types of

Salmonella plasmids that vary in size from 2 to more than 200 kb, which have been identified and characterized. Some of these include the following:

1. Virulence plasmids: These plasmids are usually large and carry genes that are responsible for the bacteria's ability to cause illness in humans and animals. These genes may encode proteins that help the bacteria evade the host immune system or enzymes that allow the bacteria to produce toxins that damage host cells [77, 78, 84, 85].

2. Antibiotic resistance plasmids: These plasmids carry genes that allow the bacteria to resist the effects of certain antibiotics. This can make the bacteria more difficult to treat and can contribute to the spread of antibiotic resistance [74, 82, 83].

3. Conjugative plasmids: These plasmids can be transferred from one bacterium to another through a process called conjugation. This allows the plasmids to spread through bacterial populations, even between species of bacteria and can contribute to the evolution of new pathogenic strains [75, 86].

4. IncI1 plasmids: These plasmids are a type of conjugative plasmid that is commonly found in *Salmonella* enterica serovar Typhimurium, a strain of *Salmonella* that is responsible for many human infections. IncI1 plasmids carry genes that encode proteins that help the bacteria colonize and survive in the host [80, 87, 88].

5. IncF plasmids: These plasmids are another type of conjugative plasmid that is found in many strains of *Salmonella*. IncF plasmids carry genes that encode proteins that help the bacteria evade the host immune system and colonize the host intestine [89].

7. Antimicrobial resistance genes in *Salmonella*

Some strains of *Salmonella* have developed resistance to certain antibiotics, which can make it more difficult to treat infections [90–92]. These are known as antibiotic-resistant *Salmonella* or AMR *Salmonella*. AMR is a growing global health concern because it can make it more difficult to effectively treat bacterial infections, including those caused by *Salmonella* [92–94]. The overuse and misuse of antibiotics are major contributing factors to the development of AMR in bacteria [93, 95]. Antibiotic resistance in *Salmonella* has a long history [96]. *Salmonella* have been known to cause illness for over a century, and antibiotics have been used to treat *Salmonella* infections since the 1940s [96, 97]. However, as with many other types of bacteria, *Salmonella* has developed resistance to many of the antibiotics that have been used for clinical treatment [98]. One of the first reported cases of antibiotic resistance in *Salmonella* was in the 1950s, when strains of *Salmonella* that were resistant to streptomycin were identified [96, 97]. Since then, *Salmonella's* resistance to other antibiotics, such as tetracycline and ampicillin, has also been reported [55, 99], and some strains are now resistant to multiple antimicrobial drugs or antibiotics.

Some common AMR genes found in *Salmonella* include the following:

1. *blaTEM* gene encodes for beta-lactamase, an enzyme that hydrolyzes beta-lactams (e.g., ampicillin, penicillins, and cephalosporins etc.; [100]).

2. *sul1* and *sul2* genes encode for sulfonamide-resistant dihydropteroate synthases, which when expressed can inactivate sulfonamide antibiotics [101–103].

3. *tetA* and *tetB* genes encode for tetracycline efflux pumps, which can pump tetracycline antibiotics out of the bacterial cell, making the bacteria resistant to these drugs [102, 104].

4. *qnr* gene encodes for quinolone resistance-determining region, which can make *Salmonella* resistant to quinolone antibiotics [92, 105, 106].

5. *mcr* gene encodes phosphoethanolamine transferase, which transfers the phosphatidylethanolamine residue to the lipid A of the cell membrane and provides resistance to colistin, last-resort antibiotics effective against multidrug-resistant *Salmonella* [107, 108].

The presence of an AMR gene does not necessarily mean that the bacterium will be resistant to the use of the antimicrobial drug [109, 110]. The ability of bacteria to survive antimicrobial treatment depends on many factors, including the specific strain of bacteria, the type and dosage of the drug, and the presence of other AMR genes [95, 110].

Recently, extensively drug-resistant (XDR) or more commonly known as multiple-drug resistant (MDR) *Salmonella* types, that is, *Salmonella* resistant to a wide range of antimicrobial drugs including many antibiotics that are typically used to treat *Salmonella* infections, have been on the rise, especially in developing countries [95, 111–113]. XDR *Salmonella* is of particular concern because it can be more difficult to treat and may lead to more severe or even fatal infections [95, 112]. XDR *Salmonella* can be transmitted through contaminated food, water, or surfaces, as well as through contact with infected animals or people. XDR phenotype in *Salmonella* arises through the acquisition of multiple AMR genes, which enables the bacteria to survive exposure to multiple drugs [109, 114]. The specific AMR genes present in XDR *Salmonella* can vary, but they may include genes that confer resistance to antibiotics such as ciprofloxacin, amoxicillin, and ceftriaxone. China has recently reported the first case of a waterborne outbreak caused by XDR *S.* Typhi in Beijing [113]. Similarly, the World Health Organization (WHO) recorded about 5274 cases of XDR typhoid fever in Pakistan from November 2016 to December 2018 [115, 116]. The prevalence of AMR in *Salmonella* can vary significantly by region, with some areas having higher rates of AMR than others. For example, studies have shown that the prevalence of AMR in *Salmonella* isolates from animals and food in the United States is generally low, with most isolates being susceptible to a range of antimicrobial drugs [7].

However, the prevalence of AMR *Salmonella* isolates from humans in the United States is higher, with some studies reporting resistance rates as high as 30–40% [93]. In other parts of the world, the prevalence of AMR *Salmonella* may be higher. For example, studies have shown that the prevalence of AMR *Salmonella* isolates from humans in some European countries is as high as 50–60% [117]. The distribution of AMR *Salmonella* in developing countries can vary significantly depending on the specific country and region. However, in general, the prevalence of AMR in *Salmonella* in developing countries tends to be higher than in developed countries [95]. There are several factors that may contribute to the higher prevalence of AMR in *Salmonella* in developing countries including the following [7]:

1. Limited access to clean water and sanitation: In some developing countries, access to clean water and adequate sanitation facilities is limited, which can increase the risk of bacterial infections, including salmonellosis and the spread of AMR.

2. Poor infection control practices: In certain countries, infection control practices are inadequate, which can increase the risk of *Salmonella* infections and the spread of AMR.

3. High use of antimicrobial drugs in animals: In some developing countries, there is high and uncontrolled use of antimicrobial drugs in animals, which contribute to the development and spread of AMR.

4. Limited surveillance and monitoring: In developing countries, there is usually limited surveillance and monitoring systems for the presence of *Salmonella* on food and food-related environment, leading to increased prevalence of infections including *Salmonella* infections and the spread of AMR *Salmonella*.

Overall, the frequency distribution of AMR in *Salmonella* among developing countries can vary significantly, but it is generally considered a major public health concern.

8. Detection and diagnosis

Accurate, sensitive, and specific detection of *Salmonella* is critical for food safety worldwide. Over the last decade, various other detection methods and techniques such as immunology, molecular biology, mass spectrometry, spectroscopy, optical analysis, and biosensor-based methods have been developed [34, 118, 119]. Generally, these methods can be divided into many categories as follows:

1. Culture methods: One of the most common methods for diagnosing *Salmonella* is through the use of traditional culture-based techniques, which are usually slow, labor-intensive, and not suitable for on-location or high-volume testing. However, these methods are considered gold standard and are in use since the discovery of enteric fever and have been standardized by the International Organization of Standards [120] for *Salmonella* detection, which is being used by many regulatory bodies all over the world [118, 121]. Similar standards have been published by FDA's Bacteriological Analytical Manual (BAM). The first stage in traditional culture methods for most food samples involves pre-enrichment in a nonselective liquid medium, such as buffered peptone water, which is then subcultured into two selective enrichment media, such as Rappaport Vasiliadis Soy broth (RVS) and Muller-Kauffmann Tetrathionate-Novobiocin (MKTTn) broth that inhibit background flora. This is followed by the inoculation on at least two selective differential agar media, such as Brilliant Green Sulfa (BGS), Bismuth Sulfite (BS), BrillianceTM *Salmonella* Agar, Xylose Lysine Deoxycholate (XLD), Xylose Lysine Tergitol-4 (XLT-4), and others, to allow the growth of *Salmonella* and distinguish them from other background microbial flora [118]. The last step in traditional culture methods includes the confirmation of presumptive positive *Salmonella* colonies.

2. Culture-independent diagnostic tests: These methods do not require prior culture enrichments in the media and can achieve sensitive and selective identification of *Salmonella*. A rapid and sensitive Whole Genome Culture-Independent Diagnostic Test (WG-CIDT) for *Salmonella* detection in lettuce has been developed, which could also be adapted for other perishable food [122].

3. Immunological assays: These include the enzyme-linked immunosorbent assay (ELISA), latex agglutination assay, and lateral flow and immunochromatography assay [118].

4. DNA detection methods: These include widely used polymerase chain reaction (PCR), mostly using invasin gene A (*invA*), loop-mediated isothermal amplification (LAMP), nucleic acid sequence-based amplification (NASBA), recombinase polymerase amplification (RPA), DNA microarrays, and others [118, 123].

5. Whole-genome sequencing methods (WGS): WGS has rapidly changed the practice of microbiology and public health surveillance and investigation of foodborne *Salmonella* illnesses, which allows to sequence and analyze the whole genome and provides a greater level of details. This method can also provide a one-step characterization of bacteria by identifying the species, serotype, genotype, and resistance and virulence genes all within a single laboratory workflow. WGS has been adopted in food safety framework in various countries, including the United States and England [124–126].

6. Mass spectrometry methods: matrix-assisted laser desorption ionization-time-of-flight mass spectrometry (MALDI-TOF MS) and liquid chromatography-mass spectrometry (LC-MS) [118].

7. Spectroscopy methods: Raman spectroscopy, near-infrared (NIR) spectroscopy, hyperspectral imaging (HSI), and optical phenotyping with light diffraction technology [118].

8. Sensor-based methods: These include electrochemical sensor- or biosensor-based technologies, which can detect *Salmonella* from as low as three colony-forming units using potentiometry, conductometry, and impedimetric techniques [119, 127]. Sensing techniques for detecting *Salmonella* in food are still in the early stages of development, but they hold promise as a way to create portable biosensing platforms. The use of nanomaterials and advanced bioreceptors makes these techniques particularly promising for future use. These methods utilize different targets for sensing *Salmonella*, including [127] single-stranded DNA/RNA-based probes for sensing of *Salmonella*, immunoglobulin-based sensing of *Salmonella*, phage-based sensing of *Salmonella*, and DNA-based biosensors for sensing *Salmonella*.

Multiple materials have been tested, which provide varying degrees of selective advantages and have their own limitations. These include magnetic nanoparticles (MNPs)-based electrochemical biosensors, carbon nanoparticles-based electrochemical biosensors, metallic nanoparticles-based electrochemical biosensors, amperometric biosensors, potentiometric biosensors, conductometric biosensors, microfluidics-based biosensing platforms, Internet of Things (IOT)-supported sensing of *Salmonella*, and clustered regularly interspaced short palindromic repeats (CRISPR)-based electrochemical sensors.

Overall, the choice of diagnostic method for *Salmonella* will depend on the specific circumstances and resources available, as well as the specific goals of the diagnosis such as identifying the specific strain of *Salmonella* or determining the severity of the infection.

9. Surveillance, prevention, and control

Many socio-economic factors contribute to the spread of *Salmonella*. The main factors are poverty and lack of education [7]. Poor environmental conditions contribute to poor hygiene, which ultimately helps spread the disease. Some *Salmonella* strains can cause serious and sometimes life-threatening infections, particularly in people with compromised immune systems. Different countries have developed regulatory framework for the testing and early detection of *Salmonella* in food. The US Centers for Disease Control and Prevention (CDC) conducts surveillance for *Salmonella* in the United States through the National *Salmonella* Surveillance System. This system tracks cases of *Salmonella* infection through laboratory testing and reporting by state health departments. CDC has developed a comprehensive national *Salmonella* surveillance program in the U.S. CDC has several systems for obtaining information about *Salmonella*, each of which has different purpose and provides information on various features of the organism's epidemiology, such as the number of outbreaks, antimicrobial-resistant infections, and subtypes. These programs include Laboratory-based Enteric Disease Surveillance (LEDS), National Notifiable Diseases Surveillance System (NNDSS), Foodborne Disease Active Surveillance Network (FoodNet), National Molecular Subtyping Network for Foodborne Disease Surveillance (PulseNet), National Antimicrobial Resistance Monitoring System—enteric bacteria (NARMS), and Foodborne Disease Outbreak Surveillance System (FDOSS; https://www.cdc.gov/*Salmonella*/reportspubs/surveillance.html). Similarly, in Canada, the surveillance of *Salmonella* is conducted by the Public Health Agency of Canada (PHAC), which monitors and tracks cases of *Salmonella* in people through its integrated *Salmonella* surveillance system, collecting data from the provinces and territories under the National Enteric Surveillance Program (NESP), FoodNet Canada, and Canadian Notifiable Disease Surveillance System (https://www.canada.ca/en/public-health/services/diseases/salmonellosis-*Salmonella*/surveillance.html). In Europe, the European Centre for Disease Prevention and Control (ECDC) and the European Food Safety Authority (EFSA) are responsible for the surveillance of *Salmonella*. European Centre for Disease Prevention and Control has framed and adopted Regulation (EC) No 2160/2003 on protecting human health against *Salmonella* and other specified foodborne zoonotic agents, with the goal of controlling *Salmonella* at every stage of food production and in animal feed to reduce its prevalence and the risk to public health (https://leap.unep.org/countries/no/national-legislation/regulation-no-1703-control-*Salmonella*-and-other-food-borne). The ECDC is responsible for the surveillance of *Salmonella* infections across the European Union (EU). The organization collects data on *Salmonella* infections in humans from the EU Member States through the EU Surveillance Network for Communicable Diseases (TESSy) system. This allows the ECDC to track the number of cases, identify outbreaks, and monitor trends in *Salmonella* infections across the EU. The EFSA, on the other hand, is responsible for food safety and conducts surveillance of

Salmonella in food products. It collects data on *Salmonella* in food from EU Member States and also conducts its own risk assessments on specific food products. The EFSA also provides scientific advice and support to the European Commission and EU Member States on food safety issues, including *Salmonella* control in the food chain. Additionally, the European Union Reference Laboratory for *Salmonella* (EURL-*Salmonella*) also plays an important role in the surveillance of *Salmonella* in Europe. This laboratory is responsible for coordinating the network of national reference laboratories for *Salmonella* and providing scientific and technical support for the detection and control of *Salmonella* in food and animal feed.

Surveillance of *Salmonella* in the developing world varies from one country to another, but in general, it is less robust and comprehensive compared with that in developed countries [7, 128, 129]. In developing countries, the surveillance of *Salmonella* is conducted by the national public health department or ministry of health [7, 128]. However, the capacity for laboratory testing and data collection is limited because of the lack of resources and infrastructure. In addition, lack of awareness and education on food safety and good hygiene practices among the population, inadequate sanitation, and poor infrastructure exacerbate the spread of *Salmonella* in developing countries. However, efforts to improve them through international collaboration and aid programs, education, and capacity building are essential to curb the spread of these bacteria. To control salmonellosis, it is important to follow good hygiene practices, such as washing hands thoroughly with soap and water before handling food and cooking food to a safe temperature to kill *Salmonella* that may be present. It is also important to store food properly and avoid cross-contamination, for example, by using separate cutting boards and utensils for raw and cooked foods. In addition to these measures, it is important to control the spread of *Salmonella* in food-producing animals, as they can be a source of contamination. To be successful, *Salmonella* control requires a focus on the sources of the organism and the means of transmission to humans, which is best achieved through the One Health approach, with adequate attention paid to animal and food sources and the environment that harbors organisms and provides avenues of transmission to humans. Specific measures such as proper animal feeding and husbandry practices, effective disinfection of animal housing and equipment, monitoring wildlife sources especially avian species, and effective food safety practices are required.

10. Conclusion

In conclusion, *Salmonella* is one of the leading causes of food poisoning in humans. It is commonly found in raw or undercooked meat, poultry, eggs, and dairy products, as well as in fruits and vegetables that have come into contact with contaminated water or soil. *Salmonella* is a pathogen of great concer, which can cause severe illness and leads to death in some cases. The virulence of *Salmonella* is determined by a variety of factors including the serotype, presence of specific virulence genes, and the host's immune response. In addition, the emergence of antibiotic-resistant *Salmonella* is a significant concern in the field of food safety. *Salmonella* can acquire AMR genes through horizontal gene transfer, and the presence of these genes makes the treatment of infections more difficult and complicated. Additionally, the emergence of extensively drug-resistant (XDR) *Salmonella* has become a great public health concern because of its ability to resist multiple classes of antibiotics, leading to a significant public health concern because of prolonged illness and increased health care costs. It is important to limit the use of antibiotics to decrease the risk of antibiotic resistance and implement strategies to prevent the spread of resistant strains of *Salmonella*, such as proper food handling and sanitation practices. It is also important to develop advanced, fast, and efficient methods to monitor the emergence of antibiotic-resistant *Salmonella* to quickly detect and respond to outbreaks.

Author details

Sohail Naushad*, Dele Ogunremi and Hongsheng Huang
Ottawa Laboratory Fallowfield, Canadian Food Inspection Agency, Ottawa, Ontario, Canada

*Address all correspondence to: sohail.naushad@inspection.gc.ca

References

[1] Barnett R. Typhoid fever. Lancet. 2016;**388**:2467

[2] Chanamé Pinedo L, Mughini-Gras L, Franz E, Hald T, Pires SM. Sources and trends of human salmonellosis in Europe, 2015-2019: An analysis of outbreak data. International Journal of Food Microbiology. 2022;**379**:109850

[3] Colin P. International symposium on Salmonella and salmonellosis. Food Microbiology. 2018;**71**:1

[4] Eng S-K, Pusparajah P, Ab Mutalib N-S, Ser H-L, Chan K-G, Lee L-H. Salmonella: A review on pathogenesis, epidemiology and antibiotic resistance. Frontiers in Life Science. 2015;**8**:284-293

[5] Littman RJ. The plague of Athens: Epidemiology and paleopathology. The Mount Sinai Journal of Medicine, New York. 2009;**76**:456-467

[6] Qamar FN, Hussain W, Qureshi S. Salmonellosis including enteric fever. Pediatric Clinics of North America. 2022;**69**:65-77

[7] WHO. *Salmonella*. World Health Organization. 2018. Available from: https://www.who.int/news-room/fact-sheets/detail/salmonella-(non-typhoidal) [Accessed: January 5, 2023]

[8] Popa GL, Papa MI. Salmonella spp. infection—A continuous threat worldwide. Germs. 2021;**11**:88-96

[9] Gryglewski RW, Chlipała M. Salmonella Typhi—Historical perspective of discovery and forgotten contribution of polish anatomopathology. Folia Medica Cracoviensia. 2020;**60**:25-32

[10] Monte DFM, Sellera FP. Salmonella. Emerging Infectious Diseases. 2020;**26**:2955

[11] Salmon DE. The discovery of the germ of swine-plague. Science (New York, N.Y.). 1884;**3**:155-158

[12] Marineli F, Tsoucalas G, Karamanou M, Androutsos G. Mary Mallon (1869-1938) and the history of typhoid fever. Annals of Gastroenterology. 2013;**26**:132-134

[13] Ferrari RG, Rosario DKA, Cunha-Neto A, Mano SB, Figueiredo EES, Conte-JuniorCA.Worldwideepidemiology of *Salmonella* Serovars in animal-based foods: A meta-analysis. Applied and Environmental Microbiology. 2019;**85**(14):e00591-e00619

[14] Majowicz SE, Musto J, Scallan E, Angulo FJ, Kirk M, O'Brien SJ, et al. The global burden of Nontyphoidal *Salmonella* gastroenteritis. Clinical Infectious Diseases. 2010;**50**:882-889

[15] CDC. *Salmonella* in United States. 2022. Available from: https://www.cdc.gov/salmonella/index.html

[16] PHAC. The Public Health Agency of Canada: Yearly food-borne illness estimates for Canada. 2022. Available from: https://www.canada.ca/en/public-health/services/food-borne-illness-canada/yearly-food-borne-illness-estimates-canada.html

[17] Tsai CN, Coombes BK. Emergence of invasive Salmonella in Africa. Nature Microbiology. 2021;**6**:273-274

[18] Popoff MY, Le Minor LE. *Salmonella*. In: Bergey's Manual of Systematics of Archaea and Bacteria. 2015. pp. 1-1

[19] Chong A, Cooper KG, Kari L, Nilsson OR, Hillman C, Fleming BA, et al. Cytosolic replication in epithelial cells fuels intestinal expansion and chronic fecal shedding of *Salmonella* Typhimurium. Cell Host & Microbe. 2021;**29**:1177-1185.e1176

[20] Lee W, Kim E, Zin H, Sung S, Woo J, Lee MJ, et al. Genomic characteristics and comparative genomics analysis of Salmonella enterica subsp. enterica serovar Thompson isolated from an outbreak in South Korea. Scientific Reports. 2022;**12**:20553

[21] McClelland M, Sanderson KE, Spieth J, Clifton SW, Latreille P, Courtney L, et al. Complete genome sequence of Salmonella enterica serovar Typhimurium LT2. Nature. 2001;**413**:852-856

[22] Naushad S, Duceppe M-O, Dupras AA, Gao R, Ogunremi D. Closed genome sequences and antimicrobial resistance profiles of eight wild bird *Salmonella* isolates obtained with MinION and Illumina MiSeq sequencing. Microbiology Resource Announcements. 2020;**9**:e00228-e00220

[23] den Bakker HC, Moreno Switt AI, Govoni G, Cummings CA, Ranieri ML, Degoricija L, et al. Genome sequencing reveals diversification of virulence factor content and possible host adaptation in distinct subpopulations of *Salmonella* enterica. BMC Genomics. 2011;**12**:425

[24] Liu X, Chen Q, Li H, Zhu C, Wu C, Wang W, et al. [effect of spvB/spvC gene on Salmonella virulence and the host immune function]. Nan fang yi ke da xue xue bao =. Journal of Southern Medical University. 2015;**35**:1649-1654

[25] Sherry NL, Horan KA, Ballard SA, Gonçalves da Silva A, Gorrie CL, Schultz MB, et al. An ISO-certified genomics workflow for identification and surveillance of antimicrobial resistance. Nature Communications. 2023;**14**:60

[26] Gao R, Huang H, Hamel J, Levesque RC, Goodridge LD, Ogunremi D. Application of a high-throughput targeted sequence AmpliSeq procedure to assess the presence and variants of virulence genes in Salmonella. Microorganisms. 2022;**10**(2):369

[27] Ilyas B, Tsai CN, Coombes BK. Evolution of Salmonella-host cell interactions through a dynamic bacterial genome. Frontiers in Cellular and Infection Microbiology. 2017;**7**:428

[28] Morgan E, Campbell JD, Rowe SC, Bispham J, Stevens MP, Bowen AJ, et al. Identification of host-specific colonization factors of Salmonella enterica serovar Typhimurium. Molecular Microbiology. 2004;**54**:994-1010

[29] Valdez Y, Ferreira RB, Finlay BB. Molecular mechanisms of *Salmonella* virulence and host resistance. Current Topics in Microbiology and Immunology. 2009;**337**:93-127

[30] Mottawea W, Duceppe MO, Dupras AA, Usongo V, Jeukens J, Freschi L, et al. Salmonella enterica prophage sequence profiles reflect genome diversity and can Be used for high discrimination subtyping. Frontiers in Microbiology. 2018;**9**:836

[31] Ong SY, Pratap CB, Wan X, Hou S, Rahman AY, Saito JA, et al. The genomic blueprint of Salmonella enterica subspecies enterica serovar Typhi P-stx-12. Standards in Genomic Sciences. 2013;**7**:483-496

[32] Bharat A, Petkau A, Avery BP, Chen JC, Folster JP, Carson CA, et al.

Correlation between Phenotypic and In Silico Detection of Antimicrobial Resistance in *Salmonella* enterica in Canada Using Staramr. Microorganisms. 2022;**10**(2):292

[33] Burnett E, Ishida M, de Janon S, Naushad S, Duceppe M-O, Gao R, et al. Whole-genome sequencing reveals the presence of the blaCTX-M-65 gene in extended-Spectrum β-lactamase-producing and multi-drug-resistant clones of *Salmonella* Serovar Infantis isolated from broiler chicken environments in the Galapagos Islands. Antibiotics (Basel, Switzerland). 2021;**10**:267

[34] Cooper AL, Low AJ, Koziol AG, Thomas MC, Leclair D, Tamber S, et al. Systematic evaluation of whole genome sequence-based predictions of Salmonella serotype and antimicrobial resistance. Frontiers in Microbiology. 2020;**11**:549

[35] McDermott PF, Tyson GH, Kabera C, Chen Y, Li C, Folster JP, et al. Whole-genome sequencing for detecting antimicrobial resistance in Nontyphoidal Salmonella. Gut Pathogens. 2016;**60**:5515-5520

[36] Campioni F, Vilela FP, Cao G, Kastanis G, dos Prazeres Rodrigues D, Costa RG, et al. Whole genome sequencing analyses revealed that Salmonella enterica serovar Dublin strains from Brazil belonged to two predominant clades. Scientific Reports. 2022;**12**:10555

[37] García-Soto S, Tomaso H, Linde J, Methner U. Epidemiological analysis of Salmonella enterica subsp. enterica Serovar Dublin in German cattle herds using whole-genome sequencing. Microbiology Spectrum. 2021;**9**:e0033221

[38] Mohammed M, Thapa S. Evaluation of WGS-subtyping methods for epidemiological surveillance of foodborne salmonellosis. One Health Outlook. 2020;**2**:13

[39] Pornsukarom S, van Vliet AHM, Thakur S. Whole genome sequencing analysis of multiple *Salmonella* serovars provides insights into phylogenetic relatedness, antimicrobial resistance, and virulence markers across humans, food animals and agriculture environmental sources. BMC Genomics. 2018;**19**:801

[40] Adeolu M, Alnajar S, Naushad S, S. Gupta R. Genome-based phylogeny and taxonomy of the 'Enterobacteriales': Proposal for enterobacterales ord. nov. divided into the families Enterobacteriaceae, Erwiniaceae fam. nov., Pectobacteriaceae fam. nov., Yersiniaceae fam. nov., Hafniaceae fam. nov., Morganellaceae fam. nov., and Budviciaceae fam. nov. International Journal of Systematic and Evolutionary Microbiology. 2016;**66**:5575-5599

[41] Tindall BJ, Grimont PAD, Garrity GM, Euzeby JP. Nomenclature and taxonomy of the genus Salmonella. International Journal of Systematic and Evolutionary Microbiology. 2005;**55**:521-524

[42] Le Minor L, Veron M, Popoff M. A proposal for *Salmonella* nomenclature. Annales de Microbiologie. 1982;**133**:245-254

[43] Brenner FW, Villar RG, Angulo FJ, Tauxe R, Swaminathan B. Salmonella nomenclature. Journal of Clinical Microbiology. 2000;**38**:2465-2467

[44] Brenner F, McWhorter-Murlin A. Identification and Serotyping of *Salmonella*. Centers for Disease Control Prevention. Altanta, GA; 1998

[45] Popoff MY. WHO Collaborating Center for Reference Research on

Salmonella: Antigenic Formulas of the *Salmonella* Serovars. Paris: Pasteur Institute; 1997. pp. 56-87

[46] Popoff MY, Bockemühl J, Gheesling LL. Supplement 2002 (no. 46) to the Kauffmann–White scheme. Research in Microbiology. 2004;**155**:568-570

[47] Bäumler AJ, Tsolis RM, Ficht TA, Adams LG. Evolution of host adaptation in Salmonella enterica. Infection and Immunity. 1998;**66**:4579-4587

[48] ECDC. ECDC and EFSA (European food safety authority and European Centre for Disease Prevention and Control): The European Union one health 2021 zoonoses report. EFSA Journal. 2022;**20**:e07666

[49] Pucciarelli MG, García-Del Portillo F. Salmonella intracellular lifestyles and their impact on host-to-host transmission. Microbiology Spectrum. 2017;**5**(4):10.1128

[50] Palmer AD, Slauch JM. Mechanisms of Salmonella pathogenesis in animal models. Human and Ecological Risk Assessment: An International Journal. 2017;**23**:1877-1892

[51] Chong A, Lee S, Yang YA, Song J. The role of typhoid toxin in Salmonella typhi virulence. The Yale Journal of Biology and Medicine. 2017;**90**:283-290

[52] Kröger C, Colgan A, Srikumar S, Händler K, Sivasankaran SK, Hammarlöf DL, et al. An infection-relevant transcriptomic compendium for *Salmonella* enterica *Serovar* Typhimurium. Cell Host & Microbe. 2013;**14**:683-695

[53] Matheson N, Kingsley RA, Sturgess K, Aliyu SH, Wain J, Dougan G, et al. Ten years experience of Salmonella infections in Cambridge, UK. The Journal of Infection. 2010;**60**:21-25

[54] Feasey NA, Dougan G, Kingsley RA, Heyderman RS, Gordon MA. Invasive non-typhoidal salmonella disease: An emerging and neglected tropical disease in Africa. Lancet. 2012;**379**:2489-2499

[55] Jajere SM. A review of Salmonella enterica with particular focus on the pathogenicity and virulence factors, host specificity and antimicrobial resistance including multidrug resistance. Veterinary World. 2019;**12**:504-521

[56] White AE, Ciampa N, Chen Y, Kirk M, Nesbitt A, Bruce BB, et al. Characteristics of *campylobacter* and *Salmonella* infections and acute gastroenteritis in older adults in Australia, Canada, and the United States. Clinical Infectious Diseases. 2019;**69**:1545-1552

[57] Canović P, Gajović O, Mijailović Z. Reiter's syndrome after Salmonella infection. Srpski Arhiv za Celokupno Lekarstvo. 2004;**132**:104-107

[58] Dworkin MS, Shoemaker PC, Goldoft MJ, Kobayashi JM. Reactive arthritis and Reiter's syndrome following an outbreak of gastroenteritis caused by *Salmonella* enteritidis. Clinical Infectious Diseases: An Official Publication of the Infectious Diseases Society of America. 2001;**33**:1010-1014

[59] Ruimin G, Linru W, Dele O. In: Microorganisms B, Miroslav SM, Abdelaziz E, editors. Virulence Determinants of Non-typhoidal Salmonellae. Rijeka: IntechOpen; 2019. p. Ch. 9

[60] Belal C, Shisan B. In: Salmonella AAB, Joshua BG, editors. Salmonella: Invasion, Evasion &

Persistence. Rijeka: IntechOpen; 2012. p. Ch. 16

[61] Dos Santos AMP, Ferrari RG, Conte-Junior CA. Virulence factors in *Salmonella* Typhimurium: The sagacity of a bacterium. Current Microbiology. 2019;**76**:762-773

[62] Galán JE, Ginocchio C, Costeas P. Molecular and functional characterization of the Salmonella invasion gene invA: Homology of InvA to members of a new protein family. Journal of Bacteriology. 1992;**174**:4338-4349

[63] Guiney DG, Fierer J. The role of the spv genes in Salmonella pathogenesis. Frontiers in Microbiology. 2011;**2**:129

[64] Röder J, Hensel M. Presence of SopE and mode of infection result in increased *Salmonella*-containing vacuole damage and cytosolic release during host cell infection by *Salmonella* enterica. Cellular Microbiology. 2020;**22**:e13155

[65] Coombes BK, Lowden MJ, Bishop JL, Wickham ME, Brown NF, Duong N, et al. SseL is a salmonella-specific translocated effector integrated into the SsrB-controlled salmonella pathogenicity island 2 type III secretion system. Infection and Immunity. 2007;**75**:574-580

[66] Geng S, Wang Y, Xue Y, Wang H, Cai Y, Zhang J, et al. The SseL protein inhibits the intracellular NF-κB pathway to enhance the virulence of Salmonella Pullorum in a chicken model. Microbial Pathogenesis. 2019;**129**:1-6

[67] Rytkönen A, Poh J, Garmendia J, Boyle C, Thompson A, Liu M, et al. SseL, a Salmonella deubiquitinase required for macrophage killing and virulence. Proceedings of the National Academy of Sciences of the United States of America. 2007;**104**:3502-3507

[68] Marcus SL, Brumell JH, Pfeifer CG, Finlay BB. Salmonella pathogenicity islands: Big virulence in small packages. Microbes and Infection. 2000;**2**:145-156

[69] Sarika K, Navneet K. In: Alexandre L, Patricia R, Manuel FC, editors. Pathogenicity Island in *Salmonella*, Salmonella spp. Rijeka: IntechOpen; 2021. p. Ch. 7

[70] Lerminiaux NA, MacKenzie KD, Cameron ADS. Salmonella Pathogenicity Island 1 (SPI-1): The evolution and stabilization of a Core genomic type three secretion system. Microorganisms. 2020;**8**:576

[71] Hensel M. Salmonella pathogenicity island 2. Molecular Microbiology. 2000;**36**:1015-1023

[72] Figueira R, Holden DW. Functions of the Salmonella pathogenicity island 2 (SPI-2) type III secretion system effectors. Microbiology (Reading, England). 2012;**158**:1147-1161

[73] Blanc-Potard AB, Solomon F, Kayser J, Groisman EA. The SPI-3 pathogenicity island of Salmonella enterica. Journal of Bacteriology. 1999;**181**:998-1004

[74] Bakkeren E, Huisman JS, Fattinger SA, Hausmann A, Furter M, Egli A, et al. Salmonella persisters promote the spread of antibiotic resistance plasmids in the gut. Nature. 2019;**573**:276-280

[75] Chen K, Dong N, Zhao S, Liu L, Li R, Xie M, et al. Identification and characterization of conjugative plasmids that encode ciprofloxacin resistance in *Salmonella*. Antimicrobial Agents and Chemotherapy. 2018;**62**(8):e00575-e00618

[76] Rychlik I, Gregorova D, Hradecka H. Distribution and function of plasmids

in Salmonella enterica. Veterinary Microbiology. 2006;**112**:1-10

[77] Silva C, Puente JL, Calva E. Salmonella virulence plasmid: Pathogenesis and ecology. Pathogens and Disease. 2017;**75**. DOI: 10.1093/femspd/ftx070

[78] Gulig PA. Virulence plasmids of Salmonella typhimurium and other salmonellae. Microbial Pathogenesis. 1990;**8**:3-11

[79] Carattoli A. Plasmid-Mediated Antimicrobial Resistance in Salmonella enterica. Molecular Biology. 2003;**5**:113-122

[80] Emond-Rheault J-G, Hamel J, Jeukens J, Freschi L, Kukavica-Ibrulj I, Boyle B, et al. The Salmonella enterica Plasmidome as a reservoir of antibiotic resistance. Microorganisms. 2020;**8**:1016

[81] Portes AB, Rodrigues G, Leitão MP, Ferrari R, Conte Junior CA, Panzenhagen P. Global distribution of plasmid-mediated colistin resistance mcr gene in Salmonella: A systematic review. Journal of Applied Microbiology. 2022;**132**:872-889

[82] Bakkeren E, Gül E, Huisman JS, Steiger Y, Rocker A, Hardt W-D, et al. Impact of horizontal gene transfer on emergence and stability of cooperative virulence in Salmonella Typhimurium. Nature Communications. 2022;**13**:1939

[83] McMillan EA, Jackson CR, Frye JG. Transferable plasmids of Salmonella enterica associated with antibiotic resistance genes. Frontiers in Microbiology. 2020;**11**:562181

[84] Ahmer BM, Tran M, Heffron F. The virulence plasmid of Salmonella typhimurium is self-transmissible. Journal of Bacteriology. 1999;**181**:1364-1368

[85] Moxley RA. Chapter 7. Enterobacteriaceae: Salmonella. In: McVey DS, Kennedy M, Chengappa MM, Wilkes R, editors. Veterinary Microbiology. 4th ed. Hoboken, NJ: Wiley-Blackwell; 2022. pp. 75-87

[86] McMillan EA, Nguyen LT, Hiott LM, Sharma P, Jackson CR, Frye JG, et al. Genomic comparison of conjugative plasmids from Salmonella enterica and Escherichia coli encoding Beta-lactamases and capable of mobilizing kanamycin resistance col-like plasmids. Microorganisms. 2021;**9**(11):2205

[87] Dorr M, Silver A, Smurlick D, Arukha A, Kariyawasam S, Oladeinde A, et al. Transferability of ESBL-encoding IncN and IncI1 plasmids among field strains of different Salmonella serovars and Escherichia coli. Journal of Global Antimicrobial Resistance. 2022;**30**:88-95

[88] Hiley L, Graham RMA, Jennison AV. Characterisation of IncI1 plasmids associated with change of phage type in isolates of *Salmonella* enterica serovar Typhimurium. BMC Microbiology. 2021;**21**:92

[89] Oluwadare M, Lee MD, Grim CJ, Lipp EK, Cheng Y, Maurer JJ. The role of the Salmonella spvB IncF plasmid and its resident entry exclusion gene traS on plasmid exclusion. Frontiers in Microbiology. 2020;**11**:949

[90] Browne AJ, Kashef Hamadani BH, Kumaran EAP, Rao P, Longbottom J, Harriss E, et al. Drug-resistant enteric fever worldwide, 1990 to 2018: A systematic review and meta-analysis. BMC Medicine. 2020;**18**:1

[91] Littmann J, Viens AM, Silva DS. The super-wicked problem of antimicrobial resistance. In: Jamrozik E, Selgelid M,

editors. Ethics and Drug Resistance: Collective Responsibility for Global Public Health. Cham: Springer International Publishing; 2020. pp. 421-443

[92] Marchello CS, Carr SD, Crump JA. A systematic review on antimicrobial resistance among *Salmonella* Typhi worldwide. The American Journal of Tropical Medicine and Hygiene. 2020;**103**:2518-2527

[93] Charani E, McKee M, Balasegaram M, Mendelson M, Singh S, Holmes AH. Global burden of antimicrobial resistance: Essential pieces of a global puzzle. Lancet. 2022;**399**:2346-2347

[94] Rodrigues GL, Panzenhagen P, Ferrari RG, Dos Santos A, Paschoalin VMF, Conte-Junior CA. Frequency of antimicrobial resistance genes in Salmonella from Brazil by in silico whole-genome sequencing analysis: An overview of the last four decades. Frontiers in Microbiology. 2020;**11**:1864

[95] WHO. Antibiotic resistance. 2020. Available from: https://www.who.int/news-room/fact-sheets/detail/antibiotic-resistance

[96] Aminov RI. A brief history of the antibiotic era: Lessons learned and challenges for the future. Frontiers in Microbiology. 2010;**1**:134

[97] Davies J, Davies D. Origins and evolution of antibiotic resistance. Microbiology and Molecular Biology Reviews: MMBR. 2010;**74**:417-433

[98] Abraham EP, Chain E. An enzyme from bacteria able to destroy penicillin. Nature. 1940;**146**:837-837

[99] Ingle DJ, Ambrose RL, Baines SL, Duchene S, Gonçalves da Silva A, Lee DYJ, et al. Evolutionary dynamics of multidrug resistant Salmonella enterica serovar 4,[5],12:i:- in Australia. Nature Communications. 2021;**12**:4786

[100] Nair DVT, Venkitanarayanan K, Kollanoor Johny A. Antibiotic-resistant Salmonella in the food supply and the potential role of antibiotic alternatives for control. Foods (Basel, Switzerland). 2018;7(10):167

[101] Mąka Ł, Maćkiw E, Ścieżyńska H, Modzelewska M, Popowska M. Resistance to sulfonamides and dissemination of sul genes among Salmonella spp. isolated from food in Poland. Foodborne Pathogens and Disease. 2015;**12**:383-389

[102] Pavelquesi SLS, de Oliveira Ferreira ACA, Rodrigues ARM, de Souza Silva CM, Orsi DC, da Silva ICR. Presence of tetracycline and sulfonamide resistance genes in *Salmonella* spp.: Literature review. Antibiotics (Basel, Switzerland). 2021;**10**(11):1314

[103] Wang X, Biswas S, Paudyal N, Pan H, Li X, Fang W, et al. Antibiotic resistance in Salmonella Typhimurium isolates recovered from the food Chain through National Antimicrobial Resistance Monitoring System between 1996 and 2016. Frontiers in Microbiology. 2019;**10**:985

[104] Adesiji YO, Deekshit VK, Karunasagar I. Antimicrobial-resistant genes associated with Salmonella spp. isolated from human, poultry, and seafood sources. Food Science & Nutrition. 2014;**2**:436-442

[105] Pribul BR, Festivo ML, de Souza MM, Rodrigues Ddos P. Characterization of quinolone resistance in *Salmonella* spp. isolates from food products and human samples in Brazil.

Brazilian Journal of Microbiology: [publication of the Brazilian Society for Microbiology]. 2016;**47**:196-201

[106] Song Q, Xu Z, Gao H, Zhang D. Overview of the development of quinolone resistance in Salmonella species in China, 2005-2016. Infection and Drug Resistance. 2018;**11**:267-274

[107] Monte DF, Nelson V, Cerdeira L, Keelara S, Greene S, Griffin D, et al. Multidrug- and colistin-resistant *Salmonella* enterica 4,[5],12:i:- sequence type 34 carrying the mcr-3.1 gene on the IncHI2 plasmid recovered from a human. Journal of Medical Microbiology. 2019;**68**:986-990

[108] Vázquez X, García V, Fernández J, Bances M, de Toro M, Ladero V, et al. Colistin resistance in monophasic isolates of *Salmonella* enterica ST34 collected from meat-derived products in Spain, with or without CMY-2 Co-production. Frontiers in Microbiology. 2021;**12**:735364

[109] Levy SB, Marshall B. Antibacterial resistance worldwide: Causes, challenges and responses. Nature Medicine. 2004;**10**:S122-S129

[110] Prestinaci F, Pezzotti P, Pantosti A. Antimicrobial resistance: A global multifaceted phenomenon. Pathogens and Global Health. 2015;**109**:309-318

[111] Mansoor H, Ahmed K, Fida S, Uzair M, Asghar A, Iqbal J. Gastrointestinal and hepatobiliary complications of extensively drug-resistant typhoid at a tertiary Care Hospital in Pakistan. Cureus. 2020;**12**:e11055

[112] Patra SD, Mohakud NK, Panda RK, Sahu BR, Suar M. Prevalence and multidrug resistance in Salmonella enterica Typhimurium: An overview in South East Asia. World Journal of Microbiology & Biotechnology. 2021;**37**:185

[113] Wang Y, Lu D, Jin Y, Wang H, Lyu B, Zhang X, et al. Extensively drug-resistant (XDR) *Salmonella* Typhi outbreak by waterborne infection—Beijing municipality, China, January-February 2022. China CDC Weekly. 2022;**4**:254-258

[114] Levy SB. Multidrug resistance—A sign of the times. The New England Journal of Medicine. 1998;**338**:1376-1378

[115] Butt MH, Saleem A, Javed SO, Ullah I, Rehman MU, Islam N, et al. Rising XDR-typhoid fever cases in Pakistan: Are we heading Back to the pre-antibiotic era? Frontiers in Public Health. 2021;**9**:794868

[116] WHO. Outbreak of extensively drug resistant (XDR) typhoid fever in Pakistan. 2018a. Available from: who.int/emergencies/disease-outbreak-news/item/27-december-2018-typhoid-pakistan-en

[117] WHO. Antimicrobial resistance surveillance in Europe 2022-2020. 2022. Available from: https://www.ecdc.europa.eu/sites/default/files/documents/ECDC-WHO-AMR-report.pdf. Report also available from: https://www.ecdc.europa.eu/en/publications-data/antimicrobial-resistance-surveillance-europe-2022-2020-data

[118] Awang MS, Bustami Y, Hamzah HH, Zambry NS, Najib MA, Khalid MF, et al. Advancement in *Salmonella* detection methods: From conventional to electrochemical-based sensing detection. Biosensors. 2021;**11**(9):346

[119] Wang M, Zhang Y, Tian F, Liu X, Du S, Ren G. Overview of Raepid

detection methods for Salmonella in foods: Progress and challenges. Foods (Basel, Switzerland). 2021;**10**(10):2402

[120] Microbiology of the food chain — Horizontal method for the detection, enumeration and serotyping of Salmonella — Part 1: Detection of Salmonella spp. 2017. ISO 6579-1. Available from: https://www.iso.org/standard/56712.html

[121] Lee K-M, Runyon M, Herrman TJ, Phillips R, Hsieh J. Review of Salmonella detection and identification methods: Aspects of rapid emergency response and food safety. Food Control. 2015;**47**:264-276

[122] Ogunremi D, Dupras AA, Naushad S, Gao R, Duceppe MO, Omidi K, et al. A new whole genome culture-independent diagnostic test (WG-CIDT) for rapid detection of Salmonella in lettuce. Frontiers in Microbiology. 2020;**11**:602

[123] Robert ST, Lily YW, Pius B, Olga VP. In: Salmonella AAB, Joshua BG, editors. Molecular Technologies for Salmonella Detection. Rijeka: IntechOpen; 2012. p. Ch. 23

[124] Brown E, Dessai U, McGarry S, Gerner-Smidt P. Use of whole-genome sequencing for food safety and public health in the United States. Foodborne Pathogens and Disease. 2019;**16**:441-450

[125] Chattaway MA, Dallman TJ, Larkin L, Nair S, McCormick J, Mikhail A, et al. The transformation of reference microbiology methods and surveillance for *Salmonella* with the use of whole genome sequencing in England and Wales. Frontiers in Public Health. 2019;**7**:317

[126] Stevens EL, Carleton HA, Beal J, Tillman GE, Lindsey RL, Lauer AC, et al. Use of whole genome sequencing by the Federal Interagency Collaboration for genomics for food and feed safety in the United States. Journal of Food Protection. 2022;**85**:755-772

[127] Mahari S, Gandhi S. Recent advances in electrochemical biosensors for the detection of salmonellosis: Current prospective and challenges. Biosensors. 2022;**12**:365

[128] Herikstad H, Motarjemi Y, Tauxe RV. *Salmonella* surveillance: A global survey of public health serotyping. Epidemiology and Infection. 2002;**129**:1-8

[129] Mogasale V, Maskery B, Ochiai RL, Lee JS, Mogasale VV, Ramani E, et al. Burden of typhoid fever in low-income and middle-income countries: A systematic, literature-based update with risk-factor adjustment. The Lancet Global Health. 2014;**2**:e570-e580

Chapter 2

Salmonellosis in Food and Companion Animals and Its Public Health Importance

Joseph K.N. Kuria

Abstract

Salmonellosis in animals is caused by typhoidal and non-typhoidal *Salmonella* organisms. Non-typhoidal salmonellosis is a zoonosis of major public health concern occasioning over 155, 000 mortalities yearly worldwide. The majority of the human infections are mainly acquired directly through consumption of contaminated foods of animal origin, particularly poultry, eggs and dairy products or consumption of contaminated fruits. Rodents and will birds are the main reservoirs of non-typhoidal salmonellosis. Salmonellosis has a great economic and health impact occasioned by the cost of surveillance, investigation, treatment, and prevention in both animals and humans. Non-typhoidal salmonellosis is further complicated by the wide host range and the emergence of multidrug resistant *Salmonella* strains due to intensification of livestock production and uncontrolled antimicrobial drug use. There is a need for more innovative prevention and control measures to safeguard losses in animals and human health. This chapter will discuss salmonellosis in food and companion animals, the public health importance, and the challenges facing its control.

Keywords: salmonellosis, animals, public health, control, salmonellosis transmission

1. Introduction

Salmonellosis is caused by bacterial species in the genus *Salmonella*, a member of the family *Enterobacteriaceae*, comprising about 63 genera. *Salmonella* has a wide host range, occurring in mammals, birds, reptiles, amphibians, fish and invertebrates. The genus has two taxonomic species, based on differences in their 16S rRNA sequence analysis, namely *Salmonella* enterica and *Salmonella bongori* [1, 2]. *S. enterica* has six subspecies namely subspecies *enterica*, *salamae*, *arizonae*, *diarizonae*, *houtenae*, and *indica* [3]. *S. enterica* subsp. *enterica* is the most common subspecies and predominantly infects warm-blooded animals. *S. bongori* species is usually found in cold-blooded animals and the environment but some are occasionally associated with human disease. *Salmonella* subspecies are further classified antigenically into serotypes, or serovars, of which there are currently close to 2700 [4]. Some *Salmonella* serovars are host adapted but the majority are not and can cause disease in a broad range of hosts.

In animals, host-adapted or typhoidal *Salmonella* serovars cause severe disease in the specific hosts, characterized by septicemia but generally pose no threat to other species including humans. The non-host -adapted serovars are generally carried asymptomatically in animals although in some cases they cause disease characterized by diarrhea. These serovars are zoonotic or potentially zoonotic. In humans, they cause non-typhoid salmonellosis (NTS) the 4th most important cause of gastroenteritis [1]. It is also one of the most important bacterial zoonotic diseases, estimated to cause, 155,000 deaths yearly worldwide [5]. Non-typhoidal salmonellosis is therefore not only a major public health concern worldwide but has great negative economic impact due to the cost of surveillance, investigation, treatment and prevention of illness [6]. It is transmitted to humans through the feco-oral route, mainly by consumption of contaminated raw or improperly cooked animal products. The major sources of infection are poultry, eggs and dairy products but contaminated fresh fruits and vegetables have also been recognized as vehicles of transmission [7, 8]. Other common sources of infection include pigs products and contact with companion animals, particularly dogs, cats and horses, as well as pet reptiles such as snakes and tortoises. Contamination of animal products with *Salmonella* can also negatively impact on food trade by limiting market access NTS organisms occur widely in the environment but the main reservoirs are rodents, reptiles and wild birds [9]. The wide host range further makes control of NTS challenging. Emergence of multidrug resistant *S. enterica* strains in animals due to misuse and over use of antimicrobial agents is an added complication. Since the majority of the human infections are acquired through the consumption of contaminated foods of animal origin, NTS from animals is likely to continue to be a threat to human health. This chapter will discuss salmonellosis in food and companion animals, the associated public health risks, the challenges facing its control and future research needs. The discussion material is derived from existing literature as well as personal experience. It is hoped the chapter will be found useful by students, researchers, practitioners and managers of animal and public health.

2. History of salmonellosis

Salmonellosis has been around for centuries. It has been determined, through recent technological development, that typhoid fever was implicated in a plague which wiped out a third of the population in the city Athens, around 430 B.C. [10]. The organism *Salmonella* is named after Daniel E. Salmon, an American veterinarian. It was named in his honor by his research assistant, Theobald Smith, who isolated the first known strain of *Salmonella* from a case of hog cholera, which he named *Salmonella choleraesuis,* in 1885 [11]. The first study of *Salmonella* in humans was conducted by Karl Joseph Eberth, a German pathologist and bacteriologist, when he described a bacillus that he suspected was the cause of typhoid. The findings were later confirmed by pathologist Georg Theodor August Gaffky and the organism name "Gaffky-Eberth bacillus", which today is known as *S. enterica* serovar Typhi [12].

Notable personalities thought to have died from *Salmonella* infection include a US president, William Henry Harrison [13] and one of the famous Wright brothers, Wilbur Wright. The importance of typhoid as a scourge in human health is exemplified by the story of "typhoid Mary "(Mary Mallon), the domestic worker who transmitted *Salmonella* to at least eight households [14].

3. Etiology of salmonellosis in animals

3.1 Classification and nomenclature

Salmonella genus is classified under the family *Enterobacteriaceae* comprising about 63 genera. Their natural habitat is the intestinal tract of animals, of which about 25, such *Shigella, Salmonella, Enterobacter, Klebsiella, Serratia, Proteus* as are clinically significant. Others, such as *Escherichia coli,* are considered part of the normal intestinal microbiota and cause disease only incidentally. Phylogenically, there are only two species in this genus, *S. enterica and bongori bongori* [1, 2], based on differences in their 16S rRNA sequence analysis. *S. enterica* has six subspecies (subspecies *enterica*, *salamae*, *arizonae*, *diarizonae*, *houtenae*, and *indica* [3].

3.2 Antigenic classification

Salmonella are further classified antigenically by the Kauffman and White classification system, which classifies the organism into serotypes on the basis of a common somatic (O), flagella (H) and capsular (K), antigens [15]. The (O) antigen is present in all serotypes and is a heat-stable component of the lipopolysaccharide (LPS) located in the outer cell membrane present in all Gram negative bacteria. The heat-labile H antigens are part of the flagella protein, flagellin, present in all motile *Salmonella* spp. Two different genes code for the flagella proteins and either or both may occur in a serovar. Only one gene is expressed at a time, hence a serovar may possess only one protein at a time and the cells are thus diphasic. The two proteins are designated as phase I and Phase II. Phase I antigens are specific to a serotype and confer serological identity whereas phase II antigens are non-specific [16]. The K antigens are heat-sensitive polysaccharides located in the bacterial capsule, which is rare among *Salmonella* serotypes. The human-restricted serovar Typhi and serovar Paratyphi C produce a variant of the K antigen, known as the virulence (Vi) antigen [17].

The antigenic structure of *Salmonella* is useful in identification of serovars. It is also a useful epidemiological tool in determining sources of infection and mode of spread [18]. In nomenclature of *Salmonella* serotypes, subspecies name is usually omitted. For instance, *S. enterica* subspecies *enterica* serotype *gallinarum* is shortened to *Salmonella* ser. *Gallinarum* or *Salmonella gallinarum* [1, 11]. There are about 2700 serotypes (serovars) so far identified [4], each having a unique combination of somatic O and flagella phase I and Phase II antigens. Over 50% of these serotypes belong to the *S. enterica* subspecies [11].

3.3 Cellular, cultural and biochemical characteristics

Salmonella species are Gram-negative non-spore forming large rods measuring 0.7–1.5 by 2.0–5.0 μm. They are motile by peritrichous flagellation with the exception of *S.* Gallinarum and *S.* Purollum. Capsulation in *salmonella* is limited to a few serovars such as *Salmonella typhi*. *Salmonella* are aerobic, facultatively anaerobic in gaseous requirements. Nutritional requirement is non-fastidious and they can be cultivated in simple media such as nutrient agar. The majority of *Salmonella* are lactose fermenters. Utilizing this characteristic, selective and differential media have been formulated for isolation, and identification. Such media include MacConkey agar, *Salmonella*-Shigella agar, brilliant green agar xylose lysine deoxycholate agar and Hektoen enteric [19].

Salmonella species have the ability of to utilize tetrathionate ($S_4O_6^{2-}$) as an alternative electron acceptor in anaerobic respiration. This confers the organism a selective growth advantage, a property that used for non-selective enrichment in cultures containing competitive bacteria [20]. Common to *Enterobacteria*, *Salmonella* are oxidase negative, catalase positive, nitrate positive and they metabolize glucose fermentative, often with gas production. Other biochemical properties used for identification of *Salmonella* include hydrogen sulfide production (except few serovars such as *Salmonella* paratyphi A, and *S. choleraesuis*), the ability to utilize citrate as a sole carbon source, decarboxylation of lysine and fermentation of dulcitol. *Salmonella* are negative for production of urease and indole, deanimation of phenylalanine or tryptophan and Voges–Proskauer reaction [21, 22].

3.4 Pathogenicity and virulence factors

Many virulence factors play a variety of roles in the pathogenesis of *Salmonella* infections. These factors enable the organism adhere to and colonize its host, invade host cell, survive and multiply in macrophages, secret toxins and evade or bypass host's defense mechanisms. The factors include capsule, flagella, fimbriae, adhesins, invasins, hemagglutinins, exotoxins and endotoxins [16]. The various virulence factors are encoded by gene clusters, referred to as *Salmonella* pathogenicity islands (SPIs), located in chromosomes, plasmids and transporons [23–25].

The polysaccharide capsular O and Vi antigens in no-typhoidal and typhoidal *Salmonella* are known to aid the organism evade host's defense by modifying the cell surface in order to inhibit host's cellular response [26, 27]. Flagella are possessed by majority of *Salmonella* serovars and are known to confer pathogenicity in addition to motility. Certain *Salmonella* serovars are able to evade or minimize the host immune response by antigenic variation of flagella antigens, from one phase to the other [24, 25]. Fimbriae are the most common adhesion factors in *Enterobacteria.* They facilitate adhesion of *Salmonella* not only to hosts' cells, thus enabling colonization, but also to surfaces and foods. They are also implicated in a variety of other roles such as biofilm formation [28], which serves to shield the organism from attack by host's defense systems.

Endotoxin or lipopolysaccharide (LPS) is located in the outer membrane of Gram negative bacteria. It is heat stable and is released only upon bacterial cell lysis. It plays a role in pathogenesis of *Salmonella* infection by evoking pyrexia, activating complement system and depressing lymphocyte function among others. Endotoxin also plays a part in septic shock that can occur in systemic infections [29, 30].

Exotoxins comprise of cytotoxins and the enterotoxins. Cytotoxins are associated with killing of the mammalian cells in vitro and probably play a role in none-secretory diarrhea [31]. There is limited information regarding the mode of action of enterotoxins of *Salmonella* but they are antigenically related to the cholera family of enterotoxins. They are associated with diarrhea disease, probably through stimulation of intestinal secretion [16].

Certain *Salmonella* strains are hemolytic, another important virulence factor mediated by HylE protein, a product of hylE gene, thought to play a role in the pathogenesis of systemic salmonellosis. The protein produces hemolysis in blood agar made from the blood of a range of animals, including humans, with certain blood types [16].

3.5 Host range

*Salmonell*a and salmonellosis occur worldwide in mammals, birds, reptiles, amphibians, fish and invertebrates. Majority of *Salmonella* are not host specific and can cause disease in a broad range of hosts but some are host restricted. *S. enterica* subsp. *enterica* is the most common and predominantly infects warm-blooded animals. In the subspecies *enterica*, serovars *typhi*, *paratyphi* and *hirschfeldii* are restricted to humans and cause typhoid and paratyphoid fever respectively. They have no significant animal or environment reservoirs. Serovars *pullorum* and *gallinarum* are restricted to poultry; *abortusovis* to sheep; *choleraesuis* to pigs; and *dublin t*o cattle [32, 33]. The rest of the serovars, referred to as non-typhoidal *Salmonella*, are zoonotic or potentially zoonotic, the most common being serovars *typhimurium* and *enteritidis* [1].

The other five *S. enterica* subspecies (*salamae, arizonae, diarizonae, houtenae* and *indica*) and *S. bongori* are usually found in cold-blooded animals and the environment but some are occasionally associated with human disease. All animal species are susceptible to *Salmonella* infection but clinical disease occurs more commonly in some and not others. Among domestic animals, poultry, cattle, pigs, poultry and horses show clinical disease but cats and dogs commonly do not [34, 35].

3.6 Isolation and identification

Isolation of *Salmonella* from samples with competing microbes involves an initial non-selective pre-enrichment followed by a selective enrichment. Selenite (SeO_3^{2-}) is inhibitory to coliforms and certain other microbial species such as fecal streptococci and is used for selective enrichment of *Salmonella spp* from both clinical and food samples. Selective enrichment is followed by plating onto selective agars, followed by biochemical and serological confirmation of suspect presumptive colonies [36]. Serogrouping by somatic and flagella antigens, can be achieved by using monovalent specific 'O', 'H' and 'Vi' antisera. Phage typing, immunomagnetic separation and ELISA-based assays are some of the screening methods developed to produce rapid results, especially from food and environmental samples [37]. Several PCR assays targeting various genes have also been developed for identification of *Salmonella.* These include the 16S rRNA, invA, agfA, viaB, hilA, sirA, ttr, bcfD and phoP genes, among others [38–40].

4. Salmonellosis in animals

*Salmonell*a has a wide host range that includes mammals, birds, reptiles, amphibians, fish and invertebrates. It is a major cause of morbidity and mortality in animals and also a major cause of economic loss in livestock. The main importance of non-typhoidal salmonellosis in animals is however its zoonosis, causing a major health, social and economic impact due to cost of surveillance, investigation and treatment. The major source of direct human infections is consumption of contaminated or infected foods of animal origin, particularly meat, eggs and dairy products, and direct contact with animals, particularly companion animals, mainly dogs, cats and horses. Rodents and wild birds are the main reservoirs of non-typhoidal salmonellosis for animals. This chapter wills therefore limit the discussion on salmonellosis to livestock food animals, companion animals, rodents and wild birds.

4.1 Salmonellosis in poultry

4.1.1 Etiology and transmission

Salmonellosis in poultry and other avian species is caused by serovars in the subspecies *enterica*. Two of the serovar, *S. pullorum* and *S. gallinarum* are avian host-specific and cause typhoidal salmonellosis, while other serovars cause non-typhoidal infections, the most important being S. *typhimurium* and *Salmonella* enteritidis [35, 41, 42] *S. enterica* subsp. *arizonae* is also recognized as a cause of paratyphoid but mainly in turkeys [43]. Although *S. pullorum* and *S. gallinarum* can infect a wide range of avian species, clinical signs are observed in a few, which include chickens, turkeys and wild birds such as quails and pheasants [42].

Transmission is horizontal via fecal-oral route and vertically via infected embrocated eggs. Transovarian infection in the egg results in subsequent infection in chicks or poults and is one of the most important modes of transmission of these two diseases. Some infected hens become asymptomatic carriers and continually transmit it to their progeny. This mode of transmission is particularly critical in hatcheries since it can result in widespread dissemination of the diseases. Transmission by cannibalism and through the respiratory tract has been reported. Humans constitute a big potential of disease introduction through mobility and duties. They can track infections on vehicles, footwear, clothing, hands and contaminated equipment. Mammals, particularly dogs, cats, rodents as well as insects can also act as mechanical transmitters [42, 44, 45].

Similar to typhoidal salmonellosis, non-typhoid salmonellosis in poultry is transmitted vertically or horizontally. *S. enteritidis* serovar has a particular preference for vertical transmission. Horizontal transmission occurs through fecal contamination of feed and drinking water and by penetration of microorganisms into the egg subsequent to fecal contamination. Infection can be introduced into a farm by humans through clothing, footwear, equipment and vehicles. Rodents and wild birds are a notable reservoir of paratyphoid *Salmonella*. They are attracted into poultry houses by left-over feed and contaminate the feed by fecal material [42, 46, 47]. Dogs and cats can also track *Salmonella* infections over long distances to contaminate farms.

4.1.2 Clinical signs

Both *S. gallinarum* and *S. pullorum* cause systemic disease but whereas the former affects birds of all ages, *S. pullorum* affects primarily young ones. Birds hatched from infected eggs may be found dead in the hatching trays. Young birds may die soon after hatching without any observable signs and most acute outbreaks occur in birds under three weeks of age. In mature birds, infection is manifested by decreased egg production, fertility, hatchability and by anorexia. Diarrhea, which is usually white or yellow, watery to mucoid, is common, with fecal pasting seen around the vent.

4.1.3 Post mortem lesions

Lesions from *S. gallinarum* and *S. pullorum* infections are characterized by septicemia, with inflammation of all internal organs, including intestines, and notably liver and spleen, which show classic gray granulomatous nodules [46]. Infected ovaries may be misshapen and/or shrunken and follicles are often pedunculated, being attached to the ovary by fibrous stalks, while the abnormal ova may contain caseous material [45]. Impaction of oviducts, resulting in egg peritonitis, is also common [48]. Recently hatched chicks show signs of septicemia and omphalitis, a condition characterized by infected yolk sacs, often accompanied by unhealed navels. The yolk sacs usually contain creamy or greenish, caseous material [49–51].

In non- typhoidal salmonellosis, the highest morbidity and death rates are usually observed during the first 2 weeks after hatching. Infected adult birds are asymptomatic and do not present signs of the disease and main importance is human infection, through consumption of contaminated meat and eggs. In young birds, it may however cause enteritis with dissemination toward the spleen, lungs, liver, spleen, and kidneys [43]. An enlarged, friable liver, with necrotic foci, is common. Chicks infected transovariary will show signs and lesions similar to those in typhoidal salmonellosis.

4.1.4 Diagnosis, treatment and control

Diagnosis of salmonellosis in poultry is achieved through significant clinical signs, necropsy finding as well as isolation and identification of the organism. *S. gallinarum* and *S. pullorum* can be differentiated with biochemical, serological tests and PCR.

S. gallinarum and *S. pullorum* may survive for a long time, months or even years in the environment, which makes it difficult to eliminate them in infected poultry houses. Once a flock is infected, the amount of *Salmonella* can be reduced, but not completely eliminated and depopulation is usually the only option. Pullorum disease and fowl typhoid are notifiable disease in many countries under OIE guidelines. Both diseases can be controlled and eradicated by use of serological testing and elimination of positive birds but vaccines may be used to control the disease. The diseases have largely been eradicated from commercial poultry in developed countries. Various antibiotics can be used to treat clinical cases, but they do not eliminate the organisms from the flock. The two serovars are highly adapted to the host species, and therefore are of little public health significance [45, 51].

4.2 Salmonellosis in cattle

4.2.1 Etiology and transmission

Salmonellosis in cattle is caused mainly by *S. enterica* ser *dublin*. The serovar is adapted to cattle but can also cause infection in other species including human. It causes economic losses in cattle production and is also a threat to human health [52]. Other serovars can also infect cattle and indeed, majority of *Salmonella* isolated from cattle are the non–host specific [53]. *Salmonella* infection in cattle is most commonly acquired by ingestion of feed or water contaminated by fecal matter from other livestock, rodents and wild birds or by contaminated animal by-products. *Salmonella* are shed by clinically infected animals and contaminate feed, water, yards, and equipment. The bacterium is also shed in saliva, nasal secretions, urine and milk in cases of systemic illness. Aerosol transmission between animals is considered possible in closely confined production systems [53–55]. Probability of vertical transmission from a dam to fetus, with calves born already infected, has been proposed [56]. The outcome of infection is determined by virulence of the serotype, dose of inoculum, degree of immunity and other stress factors.

4.2.2 Clinical signs

Salmonellosis in cattle affects all age groups causing both intestinal and systemic infection but is most severe in the young. Clinical presentations are highly variable and the differential diagnosis list is considerable [53]. Acute disease is characterized by fever, anorexia and diarrhea of varying degree. The feces may be foul smelling, and may contain varying amounts of blood, mucus, and shreds of intestinal lining. Lactation drops suddenly in dairy cows. Clinical signs may last up to a week and death is due to dehydration and toxemia. In newborn calves, the disease most commonly affects those that receive inadequate or no colostrum and signs may include central nervous system (CNS) signs or pneumonia, and death may occur in 1–2 days. Those calves that survive longer may develop complications such as polyarthritis, or gangrene of the extremities of limbs, ears and tail [57]. Pregnant cows may abort, either with or without other clinical signs [54, 57, 58]. Subacute disease is seen mainly in adult animals and signs may include mild fever, anorexia, diarrhea dehydration and weight loss. Chronic disease in manifested by low intermittent fever and anorexia. There may be watery diarrhea resulting in progressive dehydration and weight loss. The feces are usually normal or contain mucus or blood. Sick cows that recover may become carriers that shed *Salmonella* for varying periods of time and cause continuous new infections in the herd [59, 60].

4.2.3 Post mortem lesions

In animals that die peracutely due to septicemia, there may be no gross lesions other than extensive submucosal and subserosal petechial hemorrhages. In acute enteritis, seen mainly in calves, the small intestines typically shows a diffuse mucoid or mucohemorrhagic enteritis and the mesenteric lymph nodes are edematous, congested and greatly enlarged [58]. In adult cattle, chronic infection is characterized muco/necrotic enteritis, especially of the ileum, caecum and colon. The wall is thickened and covered with yellow-gray necrotic material overlying a red, granular surface. Characteristic "button" ulcers may be seen in the colon [61] and the mesenteric lymph nodes and spleen may be enlarged.

4.2.4 Diagnosis, treatment and control

Clinical signs of salmonellosis are indicative of infection but definitive diagnosis of infection involves isolation and identification of the organism. Response to antibiotic treatment is usually poor. Animals that recover from infection can remain carriers and shed bacteria intermittently or continuously for long, especially during stress periods such as transportation or calving. The carrier status can even progress to full blown clinical disease.

Control involves sourcing animals from disease-free herds in order to ensure a clean herd. New animals should be put on quarantine for at least 4 weeks. Continuous serological tests and fecal culture is recommended and positive animals culled. Control of rodents and wild birds, particularly in feeding troughs, is important. Routine disinfection of premises should be considered and aborting animals should be considered suspect, isolated, tested and culled. Vaccines are available as part of a prevention or control tool.

DOI: http://dx.doi.org/10.5772/intechopen.109324

4.3 Salmonellosis in pigs

4.3.1 Etiology and transmission

Salmonellosis in swine occurs in form of two clinical disease entities, typhoidal and non-typhoidal. Typhoidal salmonellosis is mainly caused *S. choleraesuis*. This serovar is adapted to swine and do not commonly affect other animals including, humans. Non-typhoidal infections are caused by *S. typhimurium* and is the most commonly found serotype in pigs and a common source of food poisoning in humans Other serotypes that commonly infect pigs are *e*nteritidis, agona, d*erby, hadar* and h*eidelberg* [62, 63].

The main route of transmission is feco-oral, which is exacerbated by poor hygiene and overstocking [45]. Pigs start shedding the bacteria shortly after infection and can continue to shed up to 5 months after recovery from the illness. Feed ingredients of animal origin, are another important source of infection for pigs. Mechanical transmission can be effected by humans through tracking of infections on vehicles, footwear, clothing, hands and contaminated equipment [64]. *Salmonella* also localizes in the tonsils and can lead to nose-to-nose transmission [65, 66]. Piglets can also get infected by the sow through milk, although rarely [67]. In addition, transmission of non-typhoidal salmonellosis can occur indirectly through contamination of feed and water by infections carried in the intestinal tract of wild birds and rodents.

4.3.2 Clinical signs

S. choleraesuis infections may occur at any age, but are more frequent in growing pigs, between 8 weeks and 5 months old. Outbreaks are frequently associated with stress conditions such as overcrowding, transportation, weather, concurrent infectious diseases such as parasitism, and poor management [68]. The disease is manifested as an acute septicemia characterized by fever, depression and anorexia. Sudden death is quite common in the acute phase of the disease, with pigs showing signs of cyanosis on the extremities such as ears, nose and tail, due to septicemia [69]. Pigs that survive the acute phase will show signs of yellow diarrhea and coughing. The diarrhea is foul- smelling and may contain blood and mucus. The bacterium may cross the blood-brain barrier during the septicemia phase and cause meningitis and nervous signs may be observed, but rarely. Arthritis may also be observed subsequent to localization of the organism in joints. Sick pregnant sows may abort.

Morbidity in *S. choleraesuis* infection is usually low (less than 10%) but mortality is high. The organism may localize in the mesenteric lymph nodes and such subclinical carriers intermittently or continuously shed the organism in feces, particularly under stress conditions [70].

Clinical signs of *S. typhimurium* are not common in well managed commercial herds but can occur in stressed and immuno-compromised ones. The main symptoms are fever, anorexia, yellowish diarrhea, dehydration, prostration, and mortality [71]. Affected pigs may recover in a period of one week but re-infection is common within the next three to four weeks. Mortality is rare, but those animals that survive can remain carriers, and therefore a source of continuous infection, for up to five months after recovery.

4.3.3 Post mortem lesions

In pigs that die suddenly from *S. choleraesuis* infection, the most common lesion is skin cyanosis, particularly on the ears, feet and abdomen, accompanied by swelling of the gallbladder, lymph nodes, spleen and liver. There may be necrotic foci in the liver, as well as icterus. Consolidative bacterial pneumonia will be observed in pigs that show coughing. In Pigs that show signs of diarrhea, intestinal lesions, mainly pseudomembranous inflammation of the ileum and button ulcers in the colon will be observed.

The most common macroscopic lesion in *S. typhimurium* infection in pigs is inflammation of the ileum, the caecum and colon. The inflammation is characterized the presence of yellowish necrotic pseudomembranes. Mesenteric lymph nodes may be inflamed and enlarged. Characteristic "button' ulcers may be observed in the colon [72]. Some cases of rectal strictures have been reported after clinical salmonellosis. In these cases, pigs cannot defecate and intestinal contents remain trapped in the intestines, creating severe distension.

4.3.4 Diagnosis, treatment, and control

Clinical signs and lesions found during necropsy can be indicative of salmonellosis but not diagnostic. A definitive diagnosis is achieved by isolation and identification of the organism from suitable samples such as lung, liver, spleen, kidney, or lymph nodes [73]. Isolation from the intestine or feces is often unsuccessful.

Clinical disease can be controlled my antimicrobial therapy early in the onset of the disease but this will not eliminate the pathogen. The prophylactic use of antimicrobial agents is also not recommended because of expense, and promotion of antimicrobial resistance. Vaccines are available for preventing infection but their efficacy is often disappointing [74]. However, good management and husbandry is the best method of preventing clinical disease. This involves, but is not limited to, proper cleaning and disinfection. All-in-all-out pig flow and rodent control should be part of management procedures.

4.4 Salmonellosis in companion animals

4.4.1 Salmonellosis in dogs and cats

4.4.1.1 Etiology and transmission

Numerous *Salmonella* serovars have been isolated from dogs and cats with *S. typhimurium* and *S. enteritidis* being the most common serovars. There are no host-adapted serovars identified in dogs or cats [75–77]. Most dogs and cats are asymptomatic carriers and prevalence of *Salmonella* in dogs is associated with raw feed diets and contaminated feed, due to indiscriminate feeding habits, including scavenging [76, 78]. Fecal shedding of *salmonellae* by dogs is also a possible source of infection for other dogs as well as humans [79]. Cats may get infection from eating birds and rodents [80].

4.4.1.2 Clinical signs

These are rare although some dogs and cats may manifest signs of septicemia, particularly in puppies and kittens or in adults stressed by debilitating concurrent diseases [76]. Acute gastroenteritis is the most common symptom. The signs include fever, anorexia, diarrhea and vomiting. The diarrhea may contain blood. Other syndromes may include pneumonia, pelvic limb paresis, or conjunctivitis. As enteritis progresses, abortion may occur in pregnant dogs and cats or they may give birth to weak puppies or kittens. Recovered animals can continue to shed the pathogen in their feces and saliva due to localization of the organism in the lymph nodes.

4.4.1.3 Postmortem lesions

Description of post mortem lesions in dogs and cats is scarce but the most common is enterocolitis [81]. Other recorded lesions include liver necrosis [82], pyonephrosis [83], cholecystitis [84], hemorrhagic gastroenteritis [85] and pneumonia [57].

4.4.1.4 Diagnosis, treatment and control

Diagnosis is based on isolation of the organism in conjunction with significant clinical signs. A diagnosis is conclusive if the organism is isolated from a normally sterile site, such as blood or synovial fluid in a live animal or from tissues samples from postmortem examination. Isolation of *Salmonella* may not necessarily be a definitive diagnosis in healthy animals.

Treatment for a *Salmonella* infection is primarily supportive, to compensate for the fluid lost through vomiting and diarrhea. Depending on the extent of the infection, antibiotics may be required for septic cases to prevent shock. Control of fecal contamination is of primary importance. Dogs and cats should be fed uncontaminated and properly cooked food.

4.4.2 Salmonellosis in horses

4.4.2.1 Etiology and transmission

Salmonella abortusequi is an equine-adapted serovar and is associated with abortion in mares, neonatal septicemia, polyarthritis and testicular lesions in males [24, 25, 86]. Infections are common in Asia and African but rare in the rest of the world [87]. However, the most common serovar isolated from horses is S. *typhimurium* [88].

Salmonella abortusequi transmission is oral or venereal. Infection may result from ingestion of feed contaminated by uterine discharges from mares that have recently aborted or from carrier mares. Transmission from stallions to mare during mating is also thought to occur [89]. The infection may localize in the uterus and cause repeated abortion or infection of subsequent foals.

Transmission of *S. typhimurium* is primarily fecal-oral. Feed, water and environment are contaminated by organism excreted through feces of sick or carrier horses, birds and rodents. Acutely ill animals excrete large amounts of bacteria. Risk factors for development of disease include stress due to transportation, overcrowding, changes in feed, intense physical activity, deprivation of feed and water and surgical treatment. Antibiotic treatment has also been found to increase risk for symptomless carriers. Another source of infection is eating manure, especially in foals.

4.4.2.2 Clinical signs

Serovar *abortusequi* primarily affects the reproductive system. In mares, the main clinical sign is abortion, with no other evidence of illness. Abortion usually occurs at about the seventh or eighth month of pregnancy. Retained placenta and metritis are common sequel of abortion. Foals from infected mares may develop an acute septicemia soon after birth while those that survive longer may develop polyarthritis. Sign of infection in the stallion include fever, swelling of the prepuce and scrotum, and arthritis. Epididymitis, orchitis and testicular atrophy are other abnormalities associated with infection [89].

Equine salmonellosis caused by *S. tyhimurium* can be asymptomatic, but is commonly associated with fever and. Diarrhea that can progress to septicemia in young animals [90, 91]. Infected foals are more prone to clinical disease than adult horses. Diarrhea, often severe and watery, is the most common symptom. Other symptoms include fever, colic and poor condition. The infection is often self-limiting but some conditions may progress to septicemia, resulting in death. Septicemia leads to polyarthritis, and/or pneumonia. Laminitis is a possible complication of salmonellosis in horses, and is attributed to bacterial endotoxins.

4.4.2.3 Post mortem findings

Necropsy findings in cases of *S. abortusequi* include placentitis manifested by edema, hemorrhages and areas of necrosis. Foals dying soon after birth will have nonspecific changes of acute septicemia. Polyarthritis is found in those dying at a later stage.

The main lesions in cases of *S. tyhimurium* infection in horses includes fibrinonecrotic or necrohemorrhagic enteritis, mainly in the large intestine (large colon and cecum) [90]. Other lesions reported are enterocolitis and meningoencephalomyelitis in foals [65].

4.4.2.4 Diagnosis, treatment and control

Salmonella abortusequi can be isolated from the placenta, uterine discharges, aborted foals, and the joints of foals with polyarthritis. Serological diagnosis is possible since a high titer of anti- *Salmonella* agglutinins develop in mares about 2 weeks after abortion. *S. tyhimurium* may be isolated from fecal material but this is not reliable due to intermittent shedding of the bacteria.

Antimicrobial drugs recommended in the treatment of salmonellosis should also be effective against *S. abortusequi* infection. However, antibiotics use may promote latent carrier state following recovery [92]. Isolation of infected mares and disposal of aborted material should be practiced to avoid spread of the infection and infected stallions should not be used for breeding. In areas where the disease is common, vaccination is also used as a control measure. The widespread use of vaccines is credited with the almost complete eradication of the disease in developed countries.

Antibiotic treatment of equine *S. tyhi*murium infection is not recommended, especially in cases of uncomplicated diarrhea, due to the risk of worsening symptoms, as a result of disruption of the normal intestinal microflora by the antibiotics. Instead, supportive treatment is recommended if necessary. A major problem in control is the long-term survival of the organism in the environment. Manure should be disposed of frequently and animals with diarrhea should be isolated. Rodents and wild birds control is advisable.

5. *Salmonella* from rodents and wild birds

"Typhimurium" comes from "murine" Latin for mouse, a rodent of the subfamily Murinae. Rodents and wild birds are the main reservoir for *Salmonella* in the environment. They carry the organism in their intestines, mostly asymptomatically, which they transmit to food animals in the farm environment [16, 93]. Rodents are attracted to feed and shelter around livestock farms, particularly in intensive production systems [94, 95]. Apart from *Salmonella*, rodents are carriers of a variety of other diseases such as leptospirosis and plague [96]. The source for infection is rodents' droppings which contaminate feed and water but mice and rats can also carry disease-causing organisms on their feet and hair [97]. Chicken can also get infection from eating dead mice and rats [94].

Salmonellosis in wild birds can be asymptomatic or it can be a fatal disease [98]. Asymptomatic birds may disseminate *Salmonella* to susceptible individuals through fecal shedding, shared environments, and via direct contact [99]. Birds can also transmit *Salmonella* to food animals with their feet [100]. Wild birds are particularly hazardous since they can transmit infections over long distances through migration. The most frequent serovar isolated from wild birds is *S. typhimurium* [101].

6. Public health importance of salmonellosis in animals

Non-typhoidal salmonellosis is one of the four major global causes of diarrheal diseases in human, alongside *E. coli*, Cholera and *Campylobacter* [1]. It is also one of the most important bacterial zoonotic diseases, estimated to cause, 155,000 deaths yearly worldwide [5]. Non-typhoidal salmonellosis in humans is therefore not only are major public health concerns worldwide but great negative economic impacts due to the cost of surveillance, investigation, treatment and prevention of illness [6]. It is transmitted from animals by the fecal-oral route in several ways:

1. Direct contact with infected animals. *Salmonella* is an occupational hazard for those working or living with animals [102]

2. Consumption of contaminated raw or undercooked animal products

3. Consumption of foodstuff cross-contaminated by contaminated animal products

4. Consumption of foodstuffs such as vegetables contaminated my fecal material or untreated manure from infected animals

6.1 Transmission from poultry

Poultry meat and eggs are the most common vehicles of salmonellosis to humans [7, 8] and *S. enteritidis* is one of the most commonly identified serovars in association with human infection [103]. Contamination of poultry products can occur at multiple points in the production chain. This includes during rearing, live birds transportation, slaughter, dressing and packaging [104]. During slaughter, fecal contamination of carcasses can occur from gut contents. In retail outlets, including butcheries and supermarkets, poultry meat can get contaminated or cross contaminate other products [105–107]. Leaking poultry packages can contaminate ready-to-eat foodstuffs in supermarket refrigerators and in the kitchen, poultry meat can cross-contaminate other foodstuffs during meal preparation [108], particularly, foodstuffs that are eaten raw such as fruits and salads. Eggs are important sources of *Salmonella* for humans. Eggs become contaminated either by fecal contamination of the eggshell or through transovarian transmission from infected hens [109, 110], and this can lead to human disease after consumption of the contaminated eggs. Another potential source of food contamination is poultry manure which can contaminate vegetables in the field [111].

6.2 Transmission from cattle, goats and sheep

Milk and dairy products are the second most important source of *Salmonella* infections for humans. Salmonellosis from dairy products is usually related to consumption of raw or inadequately pasteurized milk although *Salmonella* may contaminate dairy products after the pasteurization process. Milk may be contaminated by cow fecal material or manure during milking. The pathogen is shed in the feces of cows and can be present in or on the udders of cows and contaminate their milk. Unpasteurised milk and products made from it such as ice-cream, cheese, milk powder and infant formulae have been associated with *Salmonella* outbreaks [112, 113]. A variety of *Salmonella* serotypes have been isolated from these products. S. *dublin*, which is highly adapted to cattle as the primary host, has been associated with systemic form of salmonellosis in humans [52].

Goat meat, mutton, beef and beef products are recognized as important sources of human salmonellosis [114–116] Infections in most cases are associated with the consumption of raw meet, contaminated cooked meat or as a result of inadequate cooking. Organs and carcasses become contaminated with intestinal contents during slaughter and this is considered one of the important sources of infection [11]. Untreated manure can also contaminate vegetables at production stage [1, 117].

6.3 Transmission from pigs

Pork is ranked as the third most common source of human salmonellosis and *Salmonella* is the most common zoonotic pathogen affecting swine associated to human gastroenteritis [118, 119]. Many *Salmonella* serotypes are present in pigs, but the most commonly associated with foodborne illness in human is *Salmonella typhimurium*. One serotype, *S. choleraesuis* is adapted to swine as the primary host but also causes severe systemic illness in man [120], although it is not commonly isolated from pork. The most common cause of infection is eating improperly prepared or stored pork products that are contaminated with *Salmonella.*

6.4 Transmission from companion animals

Close contact between dogs and cats and their owners or those working with dogs can also be a potential source of *Salmonella* infections for humans [121, 122]. Organisms shed in the animal's feces can contaminate human food or hands. *Salmonella* shedding by dogs and cats has been incriminated in infections in humans living in the same household with the shedding pet, with children accounting for a high proportion of cases. Other persons that are particularly vulnerable are the aged and the immuno-compromised. Transmission of *Salmonella* from horses to humans in contact has also been documented [123].

7. Conclusion

Microorganism will always be with us [124], and in absence of effective control, salmonellosis in animals will continue to be a major economic and public health concern for several reasons:

1. Increased intensification of livestock production will enhance animal to animal and animal to human contact and facilitate transmission of *Salmonella* from animal to animal and from animal to human.

2. Challenges of biosecurity in intensive production systems.

3. Increased commercialization of animal food processing and marketing which will enhance food contamination and transmission to humans.

4. Zoonotic *Salmonella* serovars have a very wide host range and therefore difficult to control.

5. Emergence of multidrug resistant *S. enterica* strains in animals due to misuse and over use of antimicrobial agents.

Control of salmonellosis must therefore be addressed from these perspectives.

Salmonellosis in farms is spread by contact between animals, from the environment and from reservoirs, particularly rodents. Since the primary infection with *Salmonella* occurs at the farm level, on-farm control of *Salmonella* is critical in reducing transmission during production, thereby minimizing contamination of meet during slaughter and processing and therefore reducing food safety risks [125]. Design and implementation of innovative biosafety practices are needed. Although cleaning and disinfection are the main hygiene practices in livestock production, they are less effective in the presence of rodents. A central part of hygiene practice should therefore include rodent control. This should include design of farm structures so as to eliminate rodent breeding sites and to prevent entry of the pests into animal houses. It has been shown that even the smallest population of rodents on farms presents a hazard [94]. Innovative, safe and efficient methods of rodents control in farm structures, including use of natural predators such as barn owls, are need.

Vaccination is the most cost-effective method for prevention and control of animal diseases and the most widely used tool in veterinary medicine. It can play an important role in prevention of salmonellosis in food animals Although vaccines against *Salmonella* in various animal species are in use worldwide, their efficacy is limited probably due to the diversity and complexity of pathogenesis of *Salmonella* infections. There is need for research into more efficacious vaccines against *Salmonella.*

The close contact between companion animals and people constitutes a risk for transmission of salmonellosis particularly for children, the aged and the immunocompromised. Studies are required to determine the extent of human salmonellosis attributable to companion animals and to identify risk factors for transmission. Sensitization of animal owners, caretakers and animal and human medical practitioners on risks associated with companion animals is important.

The wide host range of NTS implies that the risk of infection for any host is high. Measures to prevent disease in animals and humans must therefore be directed at all *Salmonella* serovars. Surveillance systems designed to map the spread and identify sources of infection, particularly in humans will be of great value in control of infections.

Salmonella is a complex genus that has evolved intricate virulence and antimicrobial resistance mechanisms and uncontrolled and indiscriminate use of antibiotics has increased the isolation frequency of *Salmonella* serovars resistant to one or more antibiotics globally [6]. Non-therapeutic use of antibiotics in farms is a threat to human and animal health since majority of the human infections are acquired through the consumption of contaminated foods of animal origin. It has been demonstrated that sub-therapeutic use of antibiotics in animals may even trigger the spread *Salmonella* infection throughout a herd [125]. Whereas the global movement toward barn of antibiotics use in animals is encouraging, one of the major causes of uncontrolled antibiotic use is the commercialization of manufacture, distribution and retail of antibiotics. A significant misuse of antibiotics in humans is therefore likely to continue in absence of stringent regulation supported by surveillance data. Ongoing research on methods of blocking development of antibiotic resistance in bacteria by preventing mutation, is encouraging.

Author details

Joseph K.N. Kuria
Faculty of Veterinary Medicine, Department of Veterinary Pathology and Microbiology, University of Nairobi, Kenya

*Address all correspondence to: jknkuria@uonbi.ac.ke

References

[1] World Health Organization (WHO). *Salmonella* (non-typhoidal). 2018. Accessed from: https://www.who.int/news-room/fact-sheets/detail/salmonella-(non-typhoidal)#:~:text=Salmonella%20is%201%20of%20the,have%20been%20identified%20to%20date

[2] Su LH, Chiu CH. Salmonella: clinical importance and evolution of nomenclature. Chang Gung Medical Journal. 2007;**30**:210-219. PMID: 17760271

[3] Shima N, Nakamura J, Saito K, Kamata Y, Nagatani K, et al. *Salmonella enterica* subspecies *arizonae* detected from bilateral pleural fluid in a patient with systemic lupus erythematosus and malignant lymphoma. Internal Medicine. 2020;**59**:1223-1226. DOI: 10.2169/internalmedicine.3982-19

[4] Ryan MP, O'Dwyer J, Adley CC. Evaluation of the complex nomenclature of the clinically and veterinary significant pathogen *salmonella*. BioMed Research International. 2017;**2017**:3782182. DOI: 10.1155/2017/3782182

[5] Ao TT, Feasey NA, Gordon MA, Keddy KH, Angulo FJ, Crump JA. Global burden of invasive nontyphoidal *salmonella* disease. Emerging Infectious Diseases. 2015;**21**:941-949. DOI: 10.3201/eid2106.140999

[6] Pui CF, Wong WC, Chai LC, Tunung R, Jeyaletchumi P, et al. Review article *salmonella*: A foodborne pathogen. International Food Research Journal. 2011;**18**:465-473

[7] Bouchrif B, Paglietti B, Murgia M, Piana A, Cohen N, Ennaji M, et al. Prevalence and antibiotic-resistance of *salmonella* isolated from food in Morocco. JIDC. 2009;**28**:35-40

[8] Marcus R, Varma JK, Medus C, et al. Re-assessment of risk factors for sporadic *salmonella* serotype Enteritidis infections: A case-control study in five FoodNet sites, 2002-2003. Epidemiology and Infection. 2007;**135**:84-92

[9] Heredia N, García S. Animals as sources of food-borne pathogens: A review. Animal Nutrition. 2018;**4**(3):250-255. DOI: 10.1016/j.aninu.2018.04.006 Epub 2018 May 4

[10] Papagrigorakis MJ, Yapijakis C, SynodinosPN,EffieBaziotopoulou-ValavaniE. DNA examination of ancient dental pulp incriminates typhoid fever as a probable cause of the plague of Athens. IJID. 2006;**10**:206-214. DOI: 10.1016/j.ijid.2005.09.001

[11] Eng SK, Pusparajah P, Ab Mutalib N-S, Leng HS, Chan KG, Lee LH. *Salmonella*: A review on pathogenesis, epidemiology and antibiotic resistance. Frontiers in Life Science. 2015;**8**:284-293. DOI: 10.1080/21553769.2015.1051243

[12] Willis C, Wilson TM, Greenwood M, Ward L. Pet reptiles associated with a case of salmonellosis in an infant were carrying multiple strains of *salmonella*. Journal of Clinical Microbiology. 2002;**40**:4802-4803

[13] McHugh J, Mackowiak PA. Death in the White house: President William Henry Harrison's atypical pneumonia. Clinical Infectious Diseases. 2014;**59**:990-995. DOI: 10.1093/cid/ciu470

[14] Marineli F, Tsoucalas G, Karamanou M, Androutsos G. Mary Mallon (1869-1938) and the history of typhoid fever. Annals of Gastroenterology. 2013;**26**:132-134

[15] Chenu JW, Cox JM, Pavic A. Classification of *salmonella enterica* serotypes from Australian poultry using repetitive sequence-based PCR. Journal of Applied Microbiology. 2011;**112**(1):185-196. DOI: 10.1111/j.1365-2672.2011.05172.x

[16] Jajere SM. A review of *salmonella enterica* with particular focus on the pathogenicity and virulence factors, host specificity and antimicrobial resistance including multidrug resistance. Veterinary World. 2019;**4**:504-521. DOI: 10.14202/vetworld.2019.504-521

[17] Neupane DP, Dulal HP, Song J. Enteric fever diagnosis: Current challenges and future directions. Pathogens. 2021;**10**:410. DOI: 10.3390/pathogens10040410

[18] Al-Khafaji NSK, Al-Bavati AMK, Al-Dahmoshi HOM. In: Lamas A, Regal P, Franco CM, editors. Virulence Factors of *Salmonella* Typhi in "*Salmonella* Spp- A Global Challenge". London, UK, London, UK: IntechOpen; 2021

[19] Park SH, Ryu S, Kang DH. Development of an improved selective and differential medium for isolation of salmonella spp. Journal of Clinical Microbiology. 2012;**50**:3222-3226. DOI: 10.1128/JCM.01228-12

[20] Price-Carter M, Tingey J, Bobik TA, Roth JR. The alternative electron acceptor tetrathionate supports B12-dependent anaerobic growth of *salmonella enterica* serovar typhimurium on ethanolamine or 1, 2-propanediol. Journal of Bacteriology. 2001;**83**:2463-2475. DOI: 10.1128/JB.183.8.2463-2475.2001

[21] Veterinary bacteriology: information about important bacteria, *Salmonella* species. https://www.vetbact.org/species/210 [Accessed: June 29, 2022]

[22] Whitman WB, Rainey F, Kämpfer P, Trujillo M, Chun J, DeVos P, et al. *Salmonella* in "Bergey's Manual of Systematics of Archaea and Bacteria: *Salmonella*". Hoboken, New Jersey: Wiley; 2015

[23] Lou L, Zhang P, Piao R, Wang Y. *Salmonella* Pathogenicity Island 1 (SPI-1) and its complex regulatory network. Frontiers in Cellular and Infection Microbiology. 2019;**9**:270. DOI: 10.3389/fcimb.2019.00270

[24] Wang M, Qazi IH, Wang L, Zhou G, Han H. *Salmonella* virulence and immune escape. Microorganisms. 2020a;**8**:407. DOI: 10.3390/microorganisms8030407

[25] Wang X, Ji Y, Su J, et al. Therapeutic efficacy of phage P_{IZ} SAE-01E2 against abortion caused by *salmonella enterica* serovar Abortusequi in mice. Applied and Environmental Microbiology. 2020b;**86**(22):e01366-20. DOI: 10.1128/AEM.01366-20

[26] Marshall JM, Gunn JS. The O-antigen capsule of *salmonella enterica* Serovar typhimurium facilitates serum resistance and surface expression of FliC. Infection and Immunity. 2015;**83**:3946-3959. DOI: 10.1128/IAI.00634-15

[27] Parween F, Yadav J, J., and Qadri, A. The virulence polysaccharide of *salmonella* Typhi suppresses activation of rho family GTPases to limit inflammatory responses from epithelial cells. Cellular Infection Microbiology. 2019;**9**:141. DOI: 10.3389/fcimb.2019.00141

[28] Jin X, Marshall JS. Mechanics of biofilms formed of bacteria with fimbriae appendages. PLoS One. 2020;**15**(12):e0243280. DOI: 10.1371/journal.pone.0243280

[29] Mahamuni PP, Patil AR, Ghosh JS. Proteolytic and lipolytic properties of

endotoxins (enterotoxins) produced by *salmonella typhi* NCIM 5255, *salmonella typhimurium* NCIM 2501 and *Shigella flexneri* NCIM 5265. International Food Research Journal. 2017;**24**:2685-2688

[30] Ogoina D. Fever, fever patterns and diseases called 'fever' – A review. Journal of Infection and Public Health. 2011;**4**:108-124. DOI: 10.1016/j.jiph.2011.05.002

[31] van Asten AJAM, van Dijk JE. Distribution of "classic" virulence factors among *salmonella* spp. FEMS Immunology and Medical Microbiology. 2005;**44**:251-259. DOI: 10.1016/j.femsim.2005.02.002

[32] Wallis T. Host-specificity of *salmonella* infections in animal species. In: Maskell D, Mastroeni P, editors. *Salmonella* Infections: Clinical, Immunological and Molecular Aspects (Advances in Molecular and Cellular Microbiology). Cambridge: Cambridge University press; 2006. pp. 57-88. DOI: 101017/CBO9780511525360.004

[33] White NJ. *Salmonella typhi* (Typhoid Fever) and *S. paratyphi* (Paratyphoid Fever). 2010. Accessed from: http://www.antimicrobe.org/b106.asp

[34] Fookes M, Schroeder GN, Langridge GC, Blondel CJ, Mammina C, et al. *Salmonella bongori* provides insights into the evolution of the salmonellae. PLoS Pathogens. 2011;**7**(8):e1002191. DOI: 10.1371/journal.ppat.1002191

[35] Spickler AR. Reptile-Associated *Salmonellosis*. 2013. Retrieved from http://www.cfsph.iastate.edu/DiseaseInfo/factsheets.php.

[36] Lee KM, Runyon M, Herrmanm TJ, Phillips R, Hsieh J. Review of *salmonella* detection and identification methods: Aspects of rapid emergency response and food safety. Food Control. 2015;**47**:264-276

[37] Food Standards Australia New Zealand (FSANZ). Agents of Foodborne Illness. 2nd ed. Canberra: Food Standards Australia New Zealand 2013; 2020 https://www.foodstandards.gov.au/publications/Documents/Salmonella%20(non-typhoidal).pdf. [Accessed: July 18, 2022

[38] Chu J, Shin J, Kang S, Shin S, Chung YJ. Rapid and sensitive detection of *salmonella* species targeting the hilA gene using a loop-mediated isothermal amplification assay. Genomics Inform. 2021;**19**(3):e30. DOI: 10.5808/gi.21048 Epub 2021 Sep 30

[39] Halatsi K, Oikonomou I, Lambiri M, Mandilara G, Vatopoulos A, Kyriacou A. PCR detection of *salmonella* spp. using primers targeting the quorum sensing gene sdiA. FEMS Microbiology Letters. 2006;**259**:201-207. DOI: doi.org/10.1111/j.1574-6968.2006.00266.x

[40] Kreitlow A, Becker A, Schotte U, Malorny B, Plötz M, Abdulmawjood A, et al. Evaluation of different target genes for the detection of *salmonella* sp. by loop-mediated isothermal amplification. Letters in AEM. 2021;**72**:420-426. DOI: doi.org/10.1111/lam.13409

[41] Barrow PA, Methner U, editors. *Salmonella* in Domestic Animals. 2nd ed. Wallingford, UK: CABI; 2013. DOI: 10.1079/9781845939021.0000

[42] Spickler AR. Fowl Typhoid and Pullorum Disease. 2019. Retrieved from http://www.cfsph.iastate.edu/DiseaseInfo/factsheets.php.

[43] Chaves Hernández AJ. Poultry and Avian diseases in "Encyclopedia of Agriculture and Food Systems". 2014. Accessed from: https://www.

sciencedirect.com/topics/veterinary-science-and-veterinary-medicine/avian-disease

[44] Basnet HB, Kwon HJ, Cho SH, et al. Reproduction of fowl typhoid by respiratory challenge with salmonella Gallinarum. Avian Diseases. 2008;**52**:156-159. DOI: 10.1637/7974-032607-Reg

[45] World Organization for Animal Health (OIE). Terrestrial Manual. Chapter 2.3. 11. Fowl typhoid and pullorum disease. 2018 https://www.woah.org/fileadmin/Home/eng/Health_standards/tahm/2.03.11_FOWL_TYPHOID.pdf

[46] Dinev, Ivan. *Salmonellosis* in "Diseases of Poultry, a Colour Atlas". UK, Leicester: Ceva Sante Animal 2m printhouse Ltd; 2007. Accessed from: https://www.thepoultrysite.com/publications/diseases-of-poultry. June 24, 2022

[47] Umali DV, Lapuz RRSP, Suzuki T, Shirota K, Katoh H. Transmission and shedding patterns of *salmonella* in naturally infected captive wild roof rats (Rattus rattus) from a salmonella-contaminated layer farm. Avian Diseases. 2012;**56**:288-294. DOI: 10.1637/9911-090411-Reg.1

[48] Sibel Y, Özlem ŞY. Bacterial diseases affecting egg production of laying hens. Dairy and Veterinary Sciences Journal. 2019;**11**:555814. DOI: 10.19080/JDVS.2019.11.555814

[49] EL-Sawah AA, AL Hussien MD, Nasef SA, El-Nahass E-S, Nayel AI. Characterization of *E. coli* and *salmonella* spp. isolates associated with omphalitis in baby chicks. JVMR. 2016;**23**:91-100

[50] Khalil SA, Einas E-S. Aerobic bacteria associated with omphalitis of chicks. AJVS. 2012;**37**:69-77

[51] Shivaprasad HL. Fowl typhoid and pullorum disease. Revue Scientifique et Technique. 2000;**19**:405-424. DOI: 10.20506/rst.19.2.1222

[52] Kudirkiene E, Sørensen G, Torpdahl M, de Knegt LV, Nielsen LR, Rattenborg E, et al. Epidemiology of *salmonella enterica* Serovar Dublin in cattle and humans in Denmark, 1996 to 2016: A retrospective whole-genome-based study. Applied and Environmental Microbiology. 2020;**86**(3):e01894-e01819. DOI: 10.1128/AEM.01894-19

[53] Holschbach CL, Peek SF. Salmonella in dairy cattle. The Veterinary Clinics of North America. Food Animal Practice. 2018;**1**:133-154. DOI: 10.1016/j.cvfa.2017.10.005 Epub 2017 Dec 8

[54] Abuelo A. Salmonella Dublin in dairy calves. 2020. Accessed from: https://www.canr.msu.edu/news/salmonella-dublin-in-dairy-calves

[55] Maunsell F, Donovan GA. Biosecurity and risk management for dairy replacements. The Veterinary Clinics of North America. Food Animal Practice. 2008;**24**:155-190. DOI: 10.1016/j.cvfa.2007.10.007

[56] Hanson DL, Loneragan GH, Brown TR, Nisbet DJ, Hume ME, Edrington TS. Evidence supporting vertical transmission of *salmonella* in dairy cattle. Epidemiology and Infection. 2016;**144**:962-967. DOI: 10.1017/S0950268815002241 Epub 2015 Sep 30

[57] Spickler A and Leedom Larson KR. Salmonellosis. 2013. Retrieved from http://www.cfsph.iastate.edu/DiseaseInfo/factsheets.php

[58] Adem J, Bushra E. Bovine salmonellosis and its public health importance: A review. Advances in life.

Science and Technology. 2016;**44**:224-7181 ISSN 2225-062X

[59] Nielsen LR, Schukken YH, Gröhn YT, Ersbøll AK. *Salmonella* Dublin infection in dairy cattle: Risk factors for becoming a carrier. Preventive Veterinary Medicine. 2004;**65**:47-62. DOI: 10.1016/j.prevetmed.2004.06.010

[60] Abuelo A. Salmonella Dublin in Dairy Calves. Michigan State University extension, Dairy. 2020. Available from: https://www.canr.msu.edu/news/salmonella-dublin-in-dairy-calves

[61] Njaa BL, Panciera RJ, Clark EG., Lamm CG. Gross lesions of alimentary disease in adult cattle. Veterinary Clinics of North America Food Animal Practice. 2012;**28**:483-513. DOI: 10.1016/j.cvfa.2012.07.009

[62] Andino A, Hanning I. *Salmonella enterica*: survival, colonization, and virulence differences among serovars. Scientific World Journal. 2015;**2015**:520179. DOI: 10.1155/2015/520179 Epub 2015 Jan 13

[63] Ferrari RG, Rosario DKA, Cunha-Neto A, Mano SB, Figueiredo EES, Conte-Junior CA. Worldwide epidemiology of salmonella Serovars in animal-based foods: A meta-analysis. Applied and Environmental Microbiology. 2019;**85**(14):e00591-e00519. DOI: 10.1128/AEM.00591-19

[64] Campos J, Mourão J, Peixe L, Antunes P. Non-typhoidal *salmonella* in the pig production chain: A comprehensive analysis of its impact on human health. Pathogens. 2019;**29**:19. DOI: 10.3390/pathogens8010019

[65] Oliveira CJ, Garcia TB, Carvalho LF, Givisiez PE. Nose-to-nose transmission of salmonella typhimurium between weaned pigs. Veterinary Microbiology. 2007;**125**:355-361. DOI: 10.1016/j.vetmic.2007.05.032

[66] Van Parys A, Boyen F, Leyman B, Verbrugghe E, Haesebrouck F, Pasmans F. Tissue-specific salmonella typhimurium gene expression during persistence in pigs. PLoS One. 2011;**6**(8):e24120. DOI: 10.1371/journal.pone.0024120 Epub 2011 Aug 24

[67] Casanova-Higes A, Marín-Alcalá CM, Andrés-Barranco S, et al. Weaned piglets: Another factor to be considered for the control of *salmonella* infection in breeding pig farms. Veterinary Research. 2019;**50**:45. DOI: 10.1186/s13567-019-0666-7

[68] Schwartz KJ. Salmonella Choleraesuis in swine in "hogs, pigs and pork". 2019. Accessed from: https://swine.extension.org/salmonella-choleraesuis-in-swine/.5/7/2022

[69] Laber KE, Whary MT, Bingel SA, Goodrich JA, Smit AC, Swindle MM. Biology and diseases of swine. Laboratory Animal Medicine. 2002;**2002**:615-673. DOI: 10.1016/B978-012263951-7/50018-1 Epub 2007 Sep 2

[70] Verbrugghe E, Boyen F, Van Parys A, et al. Stress induced *salmonella* typhimurium recrudescence in pigs coincides with cortisol induced increased intracellular proliferation in macrophages. Veterinary Research. 2011;**42**(1):118. DOI: 10.1186/1297-9716-42-118

[71] Côté S, Letellier A, Lessard L, Quessy S. Distribution of *salmonella* in tissues following natural and experimental infection in pigs. Canadian Journal of Veterinary Research. 2004;**68**(4):241-248

[72] Robbins RC, Almond G, Byers E. Swine diseases and disorders.

Encyclopedia of Agriculture and Food Systems. 2014;**2014**:261-276. DOI: 10.1016/B978-0-444-52512-3.00134-0 Epub 2014 Aug 21

[73] Ramirez A. Laboratory diagnostics: *Salmonella*. 2022. Accessed from: https://www.pig333.com/articles/diagnostic-tests-for-salmonella-in-pigs_17914/

[74] Wales AD, Davies RH. Salmonella Vaccination in Pigs: A Review. Zoonoses and Public Health. 2017;**64**(1):1-13. DOI: 10.1111/zph.12256 Epub 2016 Feb 8

[75] Bataller E, García-Romero E, Llobat L, Lizana V, Jiménez-Trigos E. Dogs as a source of *salmonella* spp. in apparently healthy dogs in the Valencia region. Could it be related with intestinal lactic acid bacteria? BMC Veterinary Research. 2020;**16**:268. DOI: 10.1186/s12917-020-02492-3

[76] Carter ME, Quinn JP. *Salmonella* Infections in Dogs and Cats. Wallingford, UK: CAB International; 2000. pp. 231-244

[77] Reimschuessel R, Grabenstein M, Jake Guag J, Nemser SM, Song K. Multilaboratory survey to evaluate *salmonella* prevalence in diarrheic and nondiarrheic dogs and cats in the United States between 2012 and 2014. Journal of Clinical Microbiology. 2017;**55**:1350-1368. DOI: 10.1128/JCM.02137-16

[78] Finley R, Ribble C, Aramini J, Vandermeer M, Popa M, Litman M, et al. The risk of salmonellae shedding by dogs fed salmonella-contaminated commercial raw food diets. The Canadian Veterinary Journal. 2007;**48**:69-75

[79] Akwuobu CA, Agbo JO, Ofukwu RA. *Salmonella* infection in clinically healthy dogs in Makurdi, Benue state, north-Central Nigeria: A potential source of infection to humans. Journal of Advanced Veterinary and Animal Research. 2018;**30**:405-409. DOI: 10.5455/javar.2018.e291

[80] Centre for Diseases Control (CDC). Healthy pets, healthy people. 2021. Accessed from: https://www.cdc.gov/healthypets/pets/cats.html#:~:text=Cats%20can%20become%20infected%20with,at%20risk%20for%20serious%20illness.

[81] Sykes JE and Marks SL. Salmonellosis in 'veterinarian key'. 2016. Accessed from: https://veteriankey.com/salmonellosis-2/

[82] Giuliano A, Meiring T, Grant AJ, Watson PJ. Acute hepatic necrosis caused by salmonella enterica serotype I 4, 5, 12: -:1, 2 in a dog. Journal of Clinical Microbiology. 2015;**53**:3674-3676. DOI: 10.1128/JCM.01256-15 Epub 2015 Aug 19

[83] Crow SE, Lauerman LH, Smith KW. Pyonephrosis associated with *salmonella* infection in a dog. Journal of the American Veterinary Medical Association. 1976;**169**:1324-1326

[84] Timbs DV, Durham PJK, Barnsley DGC. Chronic cholecystitis in a dog infected with *salmonella typhimurium*. New Zealand Veterinary Journal. 1974;**22**:100-102. DOI: 10.1080/00480169.1974.34142

[85] Choudhary SP, Kalimuddin M, Prasad G, Verma BB, Narayan KG. Observations on natural and experimental salmonellosis in dogs. Journal of Diarrhoeal Diseases Research. 1985;**3**:149-153 http://www.jstor.org/stable/23497937

[86] Mahmoud MAM, Megahed G, Yousef MS, Ali FAZ, Zaki RS, AbdelhafeezHH. *Salmonella* typhimurium triggered unilateral epididymo-orchitis and splenomegaly in a Holstein bull in Assiut, Egypt: A case report.

Pathogens. 2020;**24**:314. DOI: 10.3390/pathogens9040314

[87] Grandolfo E, Parisi A, Ricci A, et al. High mortality in foals associated with *salmonella enterica subsp. enterica* Abortusequi infection in Italy. Journal of Veterinary Diagnostic Investigation. 2018;**30**:483-485. DOI: 10.1177/1040638717753965 Epub 2018 Jan 11

[88] Stewart AJ. Salmonellosis in horses in "MSD veterinary manual".2016. From https://www.msdvetmanual.com/digestive-system/intestinal-diseases-in-horses-and-foals/salmonellosis-in-horses [Accessed: September 20, 2022]

[89] Constable P, Hinchcliff K, Done S, and Grünberg W, editors. Abortion in mares and septicemia in foals associated with *Salmonella Abortusequi* (*Abortivoequina*) (Equine Paratyphoid). In: veterinary medicine. 11th ed. 2017. Accessed from: https://www.sciencedirect.com/science/article/pii/B9780702052460000188#s0935

[90] Juffo GD, Bassuino DM, Gomes DC, et al. Wurster F, Pissetti C, Pavarini SP, Driemeier D. equine salmonellosis in southern Brazil. Tropical Animal Health and Production. 2017;**49**:475-482. DOI: 10.1007/s11250-016-1216-1 Epub 2016 Dec 24

[91] Smith BP. Equine salmonellosis: A contemporary view. Equine Veterinary Journal. 1981;**13**:147-151 Available from: https://onlinelibrary.wiley.com/doi/pdf/10.1111/j.2042-3306.1981.tb03471.x

[92] Fakour, S.H. ., Musavi Rad, A. and Ahmadi, E., A. 2020. A study on latent equine salmonellosis based on phenotypic and molecular methods in Kurdistan Province of Iran, Iran Journal of Veterinary Medicine 14: 352-361. DIO 10.22059/IJVM.2020.296678.1005058

[93] Skov MN, Madsen JJ, Rahbek C, Lodal J, Jespersen JB, Jørgensen JC, et al. Transmission of salmonella between wildlife and meat-production animals in Denmark. Journal of Applied Microbiology. 2008;**105**:1558-1568. DOI: 10.1111/j.1365-2672.2008.03914.x.Pages 1558-1568

[94] Meerburg BG, Kijlstra A. Review role of rodents in transmission of salmonella and campylobacter. Journal of the Science of Food and Agriculture. 2007;**87**:2774-2781

[95] Surgeoner GA. Rodent control in livestock and poultry facilities. 2007. Accessed from: https://www.thepoultrysite.com/articles/rodent-control-in-livestock-and-poultry-facilities

[96] Backhans A, Fellström C. Rodents on pig and chicken farms - a potential threat to human and animal health. Infection Ecology & Epidemiology. 2012;**2012**:2. DOI: 10.3402/iee.v2i0.17093 Epub 2012 Feb 17

[97] Kordiyeh H. How do rodents play role in transmission of foodborne diseases? Nutrition & Food Science International Journal. 2018;**6**(1):555683. DOI: 10.19080/NFSIJ.2018.06.555683

[98] Tizard I. Salmonellosis in wild birds. Seminars in Avian and Exotic Pet Medicine. 2004;**13**:50-66

[99] Afema JA, Sischo WM. *Salmonella* in wild birds utilizing protected and human impacted habitats in Uganda. EcoHealth. 2016;**13**:558-569. DOI: 10.1007/s10393-016-1149-1

[100] Daniels MJ, Hutchings MR, Greig A. The risk of disease transmission to livestock posed by contamination of farm stored feed by wildlife excreta. Epidemiology and Infection. 2003;**130**:561-568

[101] Dos Santos EJE, Azevedo RP, Lopes ATS, et al. *Salmonella* spp. in wild free-living birds from Atlantic forest fragments in southern Bahia, Brazil. BioMed Research International. 2020;**2020**:7594136. DOI: 10.1155/2020/7594136

[102] Hoelzer K, Moreno Switt AI, Wiedmann M. Animal contact as a source of human non-typhoidal salmonellosis. Veterinary Research. 2011;**42**:34. DOI: 10.1186/1297-9716-42-34

[103] Yang Y, Ricke SC, Tellez G, Kwon YM. Quantitative tracking of *salmonella Enteritidis* transmission routes using barcode-tagged isogenic strains in chickens: Proof-of-concept study. Frontiers in Veterinary Science. 2017;**4**:15

[104] Corry JE, Allen VM, Hudson WR, Breslin MF, Davies RH. Sources of *salmonella* on broiler carcasses during transportation and processing: Modes of contamination and methods of control. Journal of Applied Microbiology. 2002;**92**:424-432. DOI: 10.1046/j.1365-2672.2002.01543.x

[105] Adeyanju GT, Ishola O. *Salmonella* and *Escherichia coli* contamination of poultry meat from a processing plant and retail markets in Ibadan, Oyo state, Nigeria. Springerplus. 2014;**12**:139. DOI: 10.1186/2193-1801-3-139

[106] Alwis US, Mudannayake DC, JayasenaDD,KamalikaJH,UbeyarathnaKJH. Evaluation of salmonella cross-contamination at retail chicken meat outlets in Kandy, Sri Lanka. Korean Journal of Agricultural Science. 2014;**41**:35-40. DOI: 10.7744/cnujas.2014.41.1.035

[107] Nidaullah H, Mohd Omar AK, Rosma A, Huda N, Sohni S. Analysis of salmonella contamination in poultry meat at various retailing, different storage temperatures and carcass cuts - A literature survey. International Journal of Poultry Science. 2016;**15**:111-120

[108] Oscar TP. Initial contamination of chicken parts with salmonella at retail and cross-contamination of cooked chicken with salmonella from raw chicken during meal preparation. Journal of Food Protection. 2013;**76**:33-39. DOI: 10.4315/0362-028X.JFP-12-224

[109] Antunes P, Mourão J, Campos J, Peixe L. Salmonellosis: The role of poultry meat. Clinical Microbiology and Infection. 2016;**22**:110-121. DOI: 10.1016/j.cmi.2015.12.004 Epub 2015 Dec 17

[110] Whiley H, Ross K. Salmonella and eggs: From production to plate. International Journal of Environmental Research and Public Health. 2015;**26**:2543-2556. DOI: 10.3390/ijerph120302543

[111] Sheffield CL, Crippen TL, Beier RC, Byrd JA. Salmonella typhimurium in chicken manure reduced or eliminated by addition of LT10002014. Journal of Applied Poultry Research. 2014;**23**:116-120. DOI: 10.3382/japr.2013-00867

[112] Lobacz A, Zulewska J. Fate of *salmonella* spp. in the fresh soft raw milk cheese during storage at different temperatures. Microorganisms. 2021;**9**:938. DOI: 10.3390/microorganisms9050938

[113] Paswan R, Park YW. Survivability of *salmonella* and *Escherichia coli* O157:H7 pathogens and food safety concerns on commercial powder milk products. Dairy. 2020;**1**:189-201. DOI: 10.3390/dairy1030014

[114] Cetin E, Temelli S, Eyigor A. Nontyphoid *salmonella* prevalence, serovar distribution and antimicrobial

resistance in slaughter sheep. Food Science of Animal Resources. 2020;**40**:21-33. DOI: 10.5851/kosfa.2019.e75

[115] Duffy L, Barlow R, Fegan N, Vanderlinde P. Prevalence and serotypes of *salmonella* associated with goats at two Australian abattoirs. Letters in Applied Microbiology. 2009;**48**:193-197. DOI: 10.1111/j.1472-765X.2008.02501.x Epub 2008 Dec 2

[116] Freitas Neto, O.C. de *et al*. Sources of human non-typhoid salmonellosis: A review. Brazilian Journal of Poultry Science (online). 2010, v. 12, n. 1 (Accessed 19 July 2022), pp. 01-11. doi: 10.1590/S1516-635X2010000100001>.

[117] Ehuwa O, Jaiswal AK, Jaiswal S. *Salmonella*, food safety and food handling practices. Food. 2021;**10**:907. DOI: 10.3390/foods10050907

[118] Argüello H, Manzanilla EG, Lynch H, Walia K, Leonard FC, Egan J, et al. Surveillance data highlights feed form, biosecurity, and disease control as significant factors associated with *salmonella* infection on farrow-to-finish pig farms. Frontiers in Microbiology. 2018;**9**:187. DOI: 10.3389/fmicb.2018.00187

[119] Evangelopoulou G, Kritas S, Govaris A, Burriel AR. Pork meat as a potential source of *salmonella enterica* subsp. *arizonae* infection in humans. Journal of Clinical Microbiology. 2014;**52**:741-744. DOI: 10.1128/JCM.02933-13 Epub 2013 Dec 11

[120] Chiu CH, Su LH, Chu C. *Salmonella enterica* serotype Choleraesuis: Epidemiology, pathogenesis, clinical disease, and treatment. Clinical Microbiology Reviews. 2004;**17**:311-322. DOI: 10.1128/CMR.17.2.311-322.2004

[121] Damborg P, Broens EM, E.M., Chomel, B.B., et al. Bacterial zoonoses transmitted by household pets: State-of-the-art and future perspectives for targeted research and policy actions. Journal of Comparative Pathology. 2016;**155**:S27-S40. DOI: 10.1016/j.jcpa.2015.03.004

[122] Wei L, Yang C, Shao W, et al. Prevalence and drug resistance of *salmonella* in dogs and cats in Xuzhou, China. Journal of Veterinary Research. 2020;**64**:263-268. DOI: 10.2478/jvetres-2020-0032

[123] Bender JB, Tsukayama DT. Horses and the risk of zoonotic infections. The Veterinary Clinics of North America. Equine Practice. 2004;**20**(3):643-653. DOI: 10.1016/j.cveq.2004.07.003

[124] Falkow S. Who speaks for the microbes? Emerging Infectious Diseases. 1998;**4**:495-497. DOI: 10.3201/eid0403.980342

[125] Food and Agricultural Organization of the Uunited Nation (FAO). Interventions for the Control of Non-typhoidal *Salmonella* spp. in Beef and Pork. Microbiological risk assessment series, 30. 2015. ISSN 1726-5274. From: https://www.fao.org/3/i5317e/I5317E.pdf [Accessed: September 20, 2022]

Chapter 3

Salmonella enterica Transmission and Antimicrobial Resistance Dynamics across One-Health Sector

Leonard I. Uzairue and Olufunke B. Shittu

Abstract

From human infection to animal production and the environment, *Salmonella* enterica has become a global-threat. The pathogen's dynamics have been determined by its transfer from sector to sector. Antibiotic-resistant bacteria can survive and proliferate in antibiotics. Misuse of antibiotics has made certain *S. enterica* resistant. The One-Health sector has antibiotic-resistant *Salmonella* (an approach that recognizes that human health is closely connected to the health of animals and the shared environment). According to certain studies, most animal and environmental *S. enterica* have virulence genes needed for human infections. *S. enterica* antibiotic resistance patterns have varied over the decades, resulting in pan-drug-resistant-strains. Plasmid-mediated fluoroquinolone resistance genes are found in One-Health *Salmonella* species. The *S. enterica* subspecies Typhi has been found to be extensively drug-resistant (XDR) in some areas. Cephalosporin-resistant *S. enterica* subspecies Typhi is a severe problem that underscores the need for Vi-conjugat-vaccines. New diagnostics for resistant-*Salmonella* in food, animal, environment, and human sectors are needed to control the spread of these deadly infections. Also, hygiene is essential as reduced transmissions have been recorded in developed countries due to improved hygienic practices. This chapter aims to discuss the transmission and antimicrobial resistance dynamics of *S. enterica* across the One-Health sector.

Keywords: *Salmonella*, transmission, one-health, resistance, detection of *Salmonella*

1. Introduction

Antibiotic-resistant *Salmonella* are pathogens that antibiotics cannot control or kill. They may survive and even increase in the presence of an antibiotic. *Salmonella* is the causative agent of salmonellosis, an intestinal illness affecting humans and animals [1]. Salmonellosis is a fairly prevalent disease that is transmitted around the world. *Salmonella* is a leading agent in the development of acute and chronic diarrhea. Some species have been linked to systemic infection and sepsis that led to the deaths of various animals and humans. Salmonellosis is significant in the One-Health strategy and is consequently of major relevance to public health [2]. Recently, resistant

Salmonella has been found in humans, animals, and the environment (One-Health sector) [3]. One-Health sector is a concept used to recognize the interrelation of human health, animal health, and the shared environment and how diseases and pathogens move across the three sectors. *Salmonella* infection has been shown to produce severe systemic illness, which is responsible for large economic losses to the commercial chicken sector due to morbidity, mortality, and decreased egg production [4, 5] as reported by the Food and Agriculture Organization (FAO). The transmission has been a subject of argument by several writers [6, 7].

The selection pressure induced by antimicrobials *Salmonella* is a driving factor behind the emergence and spread of resistant bacteria, including *Salmonella enterica* pathogens, which were genetically encoded, transmitted by successive offspring, and in some instances could be transferred horizontally to distantly related bacteria [4]. Also, employing antimicrobials in food animal husbandry has increased the emergence of resistant *S. enterica* from food-producing animals [8]. Antibiotic-resistant *Salmonella* infections have grown in recent decades, making treatment more challenging. Because antimicrobial resistance is passed from one generation of bacteria through vertical transmission, resistant bacteria, in this case *S. enterica*, keep striving. Antibiotic-resistant *Salmonella* has risen for numerous causes. *Salmonella* infections require new antibacterial classes [9]. Some scientists hypothesize that *Salmonella* antimicrobial resistance is linked to *invA* expression and other mechanisms by which Gram-negative bacteria develop resistance [10, 11]. Some *S. enterica* subspecies Typhi strains with reduced ciprofloxacin sensitivity have emerged in the Indian subcontinent, southern Asia, and sub-Saharan Africa, leading to treatment failure [9, 12, 13]. *S. enterica* subspecies Typhi has resistant to first-line antibiotics such as chloramphenicol, ampicillin, and trimethoprim-sulfamethoxazole [13, 14]. Ceftriaxone, cefotaxime, and cefixime are also used to treat enteric fever, including nalidixic- and fluoroquinolone-resistant forms [10, 11].

Continuous abuse regarding the overuse of fluoroquinolone and certain cephalosporins in the management of *Salmonella* infection is underscored by the lack of effective antimicrobial stewardship programs, which has impacted antimicrobial resistance issues. Horizontal transfer of resistance genes via genetic elements like plasmids, transposons, and integrons has also impacted antimicrobial resistance issues, rendering those previously susceptible to becoming non-susceptible. Thus, the observed resistance or reduced susceptibility of *Salmonellae* to fluoroquinolones and some cephalosporins could result from genetic modification due to gene transfer. The resistant genes in *Salmonella* are embedded in the *Salmonella* pathogenic islands (SPIs). Studies have identified several SPIs, particularly the presence of mobile genetic elements (MGEs), which caused the rapid spread of resistant genes due to the high transmissible MCEs from one bacterium to the other [15, 16]. *S. enterica* are highly associated with multiple MGEs; these MGEs are in the SPIs, which are the center of virulence of *S. enterica* [17]. This book chapter aims to discuss the transmission dynamic of antibiotic-resistant *S. enterica* across humans, environments, and animals. The chapter discusses the genus *Salmonella*; the host adaptability of *Salmonella*; the virulence determinants of *Salmonella*; the transmission of antibiotic-resistant *S. enterica* in humans, animals, and the environment; and the detection of *S. enterica* and its antimicrobial resistance.

2. The genus *Salmonella*

Dr. Daniel Salmon, a veterinary bacteriologist who worked for the United States Department of Agriculture (USDA), was honored by having his name bestowed on the genus *Salmonella* [18]. *Salmonella* is non-sporulating short Gram-negative bacilli [19], and most move with the help of peritrichous flagella. However, some serotypes of *Salmonella*, such as *Salmonella pullorum* and *Salmonella gallinarum*, are not motile [20]. They may be aerobic or facultative anaerobic, and the optimal temperature range for growth is between 5 and 45 degrees Celsius. At a temperature of 37 degrees Celsius, growth is most optimal. The optimum pH for reproduction is 7; however, *Salmonella* may live in environments with pH values ranging from 4 to 9 [21]. They grow in culture media designed for *enterobacteriaceae* and in blood agar. In addition, they grow in specialized media such as *Salmonella*-Shigella Agar and some ChromoAgar. Their colonies range from 2 to 4 millimeters in diameter and have smooth and round edges. They are slightly raised in a medium that contains carbon and nitrogen [22, 23]. When preserved in variable media such as peptone broth, colonies have the potential to maintain their viability for a significant amount of time [23, 24]. *Salmonella* strains have the biochemical capacity to catabolize nutrients, D-glucose, and other carbohydrates, except for lactose and sucrose, resulting in the generation of acid and gas [25]. *Salmonella* can utilize citrate as their only source of carbon, reduce nitrate to nitrite, and have the potential to produce hydrogen sulfide [19]. They are catalase positive and oxidase negative. They neither ferment malonate nor hydrolyze urea, and they do not produce indole. The bacterium has a coating of mucus around it, which helps to protect it from being digested by phagocytes, and it also has a fringe of fimbria placed around its outer surface, which helps it adhere to cells [19, 25, 26].

Salmonella is a member of the family of bacteria known as *Enterobacteriaceae,* which, along with other major pathogens in this group, are frequently implicated in causing illness in the small intestine. However, once the bacteria establish a foothold in the small intestine, they can move throughout the body and cause full-blown systemic disease [25]. *S. enterica* and *Salmonella bongori* are the two taxonomic species that make up the genus *Salmonella*. There are six subspecies of *S. enterica*: *S. enterica* subspecies *enterica*, *S. enterica* subspecies salamae (II), *S. enterica* subspecies arizonae (IIIa), *S. enterica* subspecies diarizonae (IIIb), *S. enterica* subspecies *houtenae* (IV), and *S. enterica* subspecies indica (VI) [25]. *S. bongori* and most of the subspecies of *S. enterica* populate the environments of cold-blooded animals, and in certain instances, *S. enterica* may cause sickness in these animals [27]. However, *S. enterica* subspecies *enterica* is the most biomedically significant subspecies. This is because these subspecies' serovars have a particularly important clinical importance in both veterinary and human disorders. Based on the structures of their flagellar (H) antigens and lipopolysaccharide (LPS) (O) antigens, *S. enterica* subspecies *enterica* may be further subdivided into approximately 2500 different serovars [25, 28].

2.1 Host adaptability of *S. enterica*

Despite their genetic connection, *Salmonella* strains may be distinguished from one another by their virulence, host adaptability, and host specificity [24, 25]. There is a wealth of epidemiologic information about *S. enterica* subspecies serovar host specificity. Certain serovars have a preference for certain hosts but are not exclusive to those hosts. Serovars Typhi, Paratyphi A, Gallinarum, and Pullorum are only transmitted from one host to another. Serovar Typhi is responsible for causing typhoid fever in humans [13, 29], while serovar Paratyphi A causes paratyphoid in humans [30]. The serovar Pullorum is responsible for causing disease in poultry, known as systemic pullorum disease [31–33], which is associated with high mortality and intestinal inflammation [34]. In contrast, serovar Gallinarum is responsible for causing severe systemic fowl [35, 36].

Both *Salmonella typhimurium* and *Salmonella enteritidis* can infect a wide variety of animal hosts. Interestingly, they are responsible for transmitting distinct illnesses in various animal species [3, 33–35]. In humans, the serovar Dublin is sometimes known to induce septicemia and gastrointestinal illness [32, 33]. The serovar Typhimurium and, less often, Enteritidis may cause enterocolitis in calves, which can lead to death from dehydration [37]. The serovars Enteritidis and Typhimurium that cause systemic illness and diarrhea in freshly born chicks are carried by older hens, who are asymptomatic carriers [4, 38]. Serovars Enteritidis and Typhimurium induce a localized, self-limiting form of enterocolitis in humans with healthy immune systems. However, immunocompromised people are more likely to develop a systemic form of the illness [29, 39]. In mouse strains that are sensitive to the illness, serovars Enteritidis and Typhimurium may induce a systemic fever similar to typhoid, although, in other investigations, they do not cause diarrhea [36, 40, 41].

2.2 Virulence determinants of *Salmonella*

Salmonella displays a wide range of virulence factors that make the bacterium harmful [9, 42]. Polymorphic surface carbohydrates, an abundance of fimbrial adhesins, phase-variable flagella, and well-structured invasion and survival mechanisms in host macrophages and other cells are likely examples of these traits [43, 44]. Nearly 200 genes on the chromosomes of *Salmonella* are crucial for the pathogenicity of the bacterium [36]. These genes are located in the five SPI-1 to SPI-5 chromosomal pathogenicity islands. A genetic component called a pathogenicity island may also be found on the chromosome [45]. This part of the chromosome exists as a separate and distinct entity from the rest of it. All pathogenicity islands have a few traits, including the inability to be identified in closely related, nonpathogenic reference species or strains and the frequent encapsulation of substantial areas of DNA (10–200 kB), containing genes that typically impart virulence to bacteria [46–51].

In most cases, they are also connected to elements like inverted repeats, transposases, and integrases [52]. Two of the type III secretion systems that allow *Salmonella* species to colonize new environments more easily are encoded by SPI-1 and SPI-2 [53]. *Salmonella* SPI-2 is only found in *Salmonella* species and is conserved across all of these species [43]. Even though SPI-2 is the sole gene unique to *Salmonella*, SPI-2 is necessary to establish bacterial invasion and internalization. At the same time, SPI-1 is necessary for developing systemic infection and intracellular replication [54–57]. The SPI-1 protein is necessary for bacterial invasion and internalization [58]. Therefore, an SPI1 gene that is both present and functioning is required for *Salmonella* species to be able to cause sickness [58]. On centisome 63 of *Salmonella* pathogenicity island-1, a 40 kb region carries a significant portion of the genes required for intestinal penetration and invasion of host cells [58]. The *Salmonella* pathogenicity island-1 includes this region [59]. Environmental isolates of *Salmonella* that had naturally occurring deletions in the SPI-1 region were unable to enter mammalian cells, according to research by Ginocchio and associates [60]. *Salmonella* isolates have the potential to colonize and infiltrate intestinal epithelial cells as well as transfer pathogenic effector proteins from the bacteria into the cytosol of the host cell due to the presence of at least 37 genes in the SPI-1. *Salmonella* isolates may also transfer harmful effector proteins into the host cell's cytoplasm. Numerous parts of the type III secretion systems (T3SSs) [61], as well as their regulators and

DOI: http://dx.doi.org/10.5772/intechopen.109229

secreted effectors, are encoded by these genes [62]. SPI-1 is included inside the *Salmonella* pathogenicity island 1. After invaders seize control of host cells, the SPI-2 genes express themselves. These genes, which are required for intracellular life, are only present in *Salmonella* for survival within epithelial cells and macrophages [63]. Mutants' pathogenicity was much diminished because they could not colonize the infected individuals' spleens and lacked SPI-2 genes [58, 63]. The effector proteins *sipA*, *sipB*, *sipC*, *sifA*, *hilA*, *hilC*, and *hilD*, as well as *invA*, *spiC*, and *invF*, are among those secreted [61]. These chromosomal clusters of virulence genes can only be found in *Salmonella* and are unique to those species.

2.3 Transmission of antibiotic-resistant *S. enterica* in humans, animals, and the environment

Antibiotic resistance mechanisms in *S. enterica* include resistance to aminoglycosides (e.g., alleles of *aacC*, *aadA*, *aadB*, *ant*, *aphA*, and *StrAB*), B-lactams (e.g., *blaCMY-2*, *TEM-1*, and *PSE-1*), chloramphenicol (e.g.,*floR*, *cmlA*, and *StrAB*), and other antibiotics [64]. In some strains of *Salmonella*, multidrug resistance mechanisms were shown to be associated with integrons or mobile genetic elements (MGEs) such as *IncA/C* plasmids [65, 66]. *Salmonella* that is resistant to antibiotics may be transmitted from animals raised for food to people; in this case, there will be similarities in the resistant patterns and genes present [67]. *Salmonella* strains that are resistant to antibiotics have been found in humans, and some of these strains have antibiotic-resistant components that are the same as those found in *Salmonella* isolated from food animals [5, 68, 69]. This suggests that these strains may have come from the same source, which is an evidence of cross-transmission.

Humans are the principal reservoir for *Salmonella* serovars Typhi and other human-specific serovars. In contrast, other animal species are the key reservoirs for non-typhoidal *Salmonella* (NTS), which has been linked to human illnesses and infections in other animal species [13]. *Salmonella* may be found in the feces of practically every animal species; as a result, the zoonotic transmission of *Salmonella* is not restricted to animals raised for human consumption alone [70, 71]. Foods produced from poultry are the primary cause of *Salmonella* infections in humans, namely, in eggs, egg products, and chicken meat. Veterinarians and public health officials have identified the shedding of *Salmonella* as a source of infections for dog handlers, dog owners, and the communities in which they live [18, 22]. This suggests that pets, and particularly dogs in close contact with humans, may be responsible for the transmission of *Salmonella*. Infected dogs may continue to be carriers of the disease and feces shedders, making them a source of *Salmonella* for humans and other animals. Although these sources are not often responsible for big outbreaks, they may be responsible for isolated occurrences [70], which is why contact with ill cattle is a systematic way for farm workers to be exposed to diseases. The Centers for Disease Control and Prevention (CDC) reported several outbreaks of multidrug-resistant *S. typhimurium* infection associated with veterinary facilities. In areas with poor sanitation and contaminated water, fecal–oral transmission from person to person is the route for enteric or typhoid fever [71]. *Salmonella typhi* is only known to be carried by humans, not any other animals. *S. enterica* serovars, which have a wide host range, are common in the populations of warm-blooded animals that contribute to the human food supply.

Bacterial transmission typically occurs through the consumption of raw or undercooked food products [63], with poultry being one of the most important reservoirs of *Salmonella* species [13, 37]. *Salmonella* strains of many different serotypes have been identified from their natural environments and food sources around the globe [29, 72]. According to Fazl and colleagues' research [45], hens are the primary vector for the vertical spread of *Salmonella*, which occurs via the ingestion of chicken eggs. *Salmonella* spreads quickly from breeding flocks to broiler and commercial egg-laying flocks. *Salmonella* spreads horizontally between birds through the fecal–oral pathway. The bacteria persist in the environment and have been isolated from poultry litter and dust [73]. The CDC reported in August 2018 about an outbreak of *Salmonella* Infantis from chicken products, which had also been reported previously [71, 74, 75].

Animal diseases are often brought on by ingesting contaminated food or water. To infiltrate the intestinal epithelium and colonize the mesenteric lymph nodes and other internal organs in the case of a systemic infection. *Salmonella* bacteria must withstand the challenging circumstances of the digestive tract [67]. Both humans and animals may get *Salmonella* infections when exposed. The ability of *Salmonella* to link with host cells and trigger its internalization has been studied [13]. These are essential for *Salmonella* to survive in the host environment and enter non-phagocytic cells. Animal waste commonly allows *Salmonella* to enter agricultural environments [76]. Plants and surface water used for irrigation or as a diluent for pesticides or fertilizers may be directly contaminated by animal feces [77]. There has been an increase in recent years in the number of reports that show a link between foodborne disease and the eating of fresh produce contaminated with *Salmonella* [78]. *Salmonella* can adapt to various environmental conditions, including those with a low pH or high temperature, allowing it to survive outside the host organism. *Salmonella* may adhere to plant surfaces and attach before actively infecting a variety of plant interiors. *Salmonella* that originates in plants retains its virulence when infecting animals [79]. Plants may thus act as a secondary host for *Salmonella* infections and contribute to spreading the bacteria to animals and humans.

2.4 Antimicrobial resistance of *S. enterica* from humans, animals, and the environment

The mechanisms of antibiotic resistance fall into three categories: (1) inactivation of the antimicrobial, (2) efflux or changes in permeability or transport of the resistance pathogen, or (3) modification or replacement of the antimicrobial target [80, 81]. Resistance is genetically encoded and may result from mutations in endogenous genes, horizontal gene transfer via plasmids, or horizontal acquisition of alien resistance genes [81, 82]. Both horizontally acquired genes and point mutations may contribute to resistance encoding. Promoter or operator point mutations might be the root cause of overexpression of endogenous genes like the *AmpC*-lactamase gene or the mar locus [83]. Some antimicrobial target genes, like the gyrase gene, are susceptible to point mutations that may turn them into resistant targets. Exogenous resistance genes encoded on plasmids, integrons, phage, and transposons can be horizontally propagated via the processes of transformation, conjugation, and transduction [84]. This includes genes that code for enzymes that render the antimicrobial inactive, such as lactamases that cleave the four-membered ring in lactams; efflux systems, such as *tet (A)*; altered versions of the enzymes the antimicrobial is intended to inhibit, such as *dfrA*; or enzymes that alter the antimicrobial target, such as ribosomal RNA methylase [85–87].

Additionally, by researching the mechanisms of resistance, one may discover the genetic link between animal and human resistance [88, 89]. Suppose the antibiotic resistances seen in human bacterial isolates are closely related to those seen in animal isolates. In that case, it may be possible to identify animal sources of resistant bacteria in human infections that can be targeted to reduce human disease [76, 90, 91]. This can be done by determining if the resistances seen in human bacterial isolates are similar to those seen in animal isolates [92]. This is possible due to the diversity of genetic factors contributing to antibiotic resistance.

Antibiotic resistance among *Salmonella* strains is increasing, which is a major cause for worry in protecting public health worldwide [93]. At the beginning of the 1960s, it was revealed that *Salmonella* had first developed resistance to a single antibiotic [88]. Since then, more *Salmonella* strains resistant to one or more antimicrobial medications have been isolated in various countries, including developed nations [94]. This trend has been seen in several countries. Traditional antibiotic therapies for *Salmonella* infections include penicillin, chloramphenicol, and trimethoprim-sulfamethoxazole, which are just a few available options. These treatments are believed to be the earliest lines of defense against *Salmonella*. *Salmonella* strains resistant to many antibiotics are referred to as multidrug-resistant *Salmonella*. The MDR phenotypic characteristic was extensively dispersed throughout *S. enterica* over an extended time, particularly in *S. typhi* and, to a lesser degree, in *Salmonella paratyphi* [68, 95]. Asia and Africa are two continents with a substantial incidence of *S. enterica* strains with the MDR feature [96]. During a surveillance investigation carried out in several nations in Asia and Africa, a significant number of *S. enterica* MDR isolates were identified [97]. The research particularly pointed to Pakistan, India, Nepal, and Vietnam, where extensive drug-resistant *Salmonella* was discovered. Because of the widespread use of fluoroquinolones and extended-spectrum cephalosporins, which were used to treat MDR *S. enterica*, there has been an increase in *S. enterica* that are capable of producing beta-lactamases [17, 98]. This is because traditional antibiotics have become less effective due to the widespread use of drugs like fluoroquinolones and extended-spectrum cephalosporins. Despite this, some evidence suggests that an increasing number of typhoid *Salmonella* are acquiring resistance to fluoroquinolones. Isolates from various nations have been reported to be resistant to nalidixic acid, which suggests that they have diminished sensitivity to ciprofloxacin and other fluoroquinolones [93, 99]. The rise in resistant non-typhoidal *Salmonella* (NTS), particularly in animals used in food production, has made controlling the spread of *S. enterica* strains resistant to antibiotics more difficult. This is true in particular for animals that are reared for their meat. According to the investigation findings, the MDR phenotype was present in most NTS clinical isolates. Public health officers have voiced their worries over treating ailment and prevention due to this phenomenon.

The use of antibiotics in animal feed to stimulate the development of food animals and in veterinary care to treat bacterial illnesses in those animals is the primary factor that contributes to the establishment of *Salmonella* with antimicrobial resistance [67]. This is a high risk of zoonotic illness due to the transfer of MDR *Salmonella* strains from animals to people via the intake of food or water contaminated with the feces of the animals, through direct contact or by the consumption of diseased food animals. Additionally, multidrug-resistant *Salmonella* strains were discovered in the aquatic habitat of some exotic pet animals, such as tortoises and turtles [100]. This might lead to an increased risk of zoonotic infections in people via direct contact with these animals [74, 76, 90].

2.5 Detection of *S. enterica* and its antimicrobial resistance

Several methods are used to detect *Salmonella* and its resistance patterns and genes. There are conventional or culture, serological, and molecular techniques, including polymerase chain reaction and sequencing. Confirming infection with *Salmonella* is required before treatment [101–103]. A diagnosis may be confirmed by culture and isolation. *Salmonella* isolates may be differentiated in various ways, and the number of *Salmonella* species is continually expanding [104, 105]. *Salmonella* is typed using complex procedures in addition to serotyping based on antigens to track individual isolates and explain pathogenicity [58, 86]. It is essential from an epidemiological standpoint to distinguish *Salmonella* isolates because definitive typing may assist in locating the source of an epidemic and tracking changes in antibiotic resistance [105].

Pre-enrichment, selective enrichment and culturing, isolation, biochemical characterization, serological characterization, and final identification are the steps that are included in the standard approach for detecting *Salmonella* [106, 107]. This method needs at least 4 days to get a negative result, and it takes between six and 7 days to identify and confirm positive samples.

Antibiotic sensitivity testing (AST) measured by inhibition zones is determined by the disk diffusion method, and it is proportional to the susceptibility of the bacteria to the antibiotic on the disk [108]. This depends on the antibiotic disk's potency and infusing ability. It may not take much modification to use disk diffusion for testing antimicrobial disks [109]. It is used to screen many isolates to choose a subset for further testing, such as MIC determinations. Antimicrobial types must include interpretation criteria (susceptible, intermediate, and resistant) based on standards, guidelines, and quality control reference organisms. Approaches to AST are selected based on their user-friendliness, versatility, adaptability to automated or semi-automated systems, cost-effectiveness, dependability, and accuracy. Conventional *Salmonella* serotyping is most typically done [110].

S. enterica serotyping is conducted on a global scale, which has enabled improvements in the monitoring and detection outbreaks on a global scale. The O (somatic), H (flagellar), and Vi (capsular) antigens from the lipopolysaccharide (LPS) layer of the cell wall are used for serotyping *Salmonella* isolates [111, 112]. *Salmonella* may spontaneously and reversibly change between these two stages of flagellar antigen synthesis, each containing a unique set of H antigens. This phenomenon is known as diphasic flagellar antigen production. In the first phase, also known as the specific phase, the various antigens are denoted by lowercase letters; in the second phase, also known as the group phase, the antigens found initially are given numbers [113, 114]. Traditional serotyping, which uses the autoagglutination method, has some important drawbacks. One is the inability to identify non-typeable Vi antigens and strains [75, 115]. It takes a lot of time, a lot of different chemicals, and a lot of experienced laboratory workers to do this [4].

Latex agglutination, enzyme immunoassay (EIA), and enzyme-linked immunosorbent assay (ELISA) are three examples of the types of immunological tests that have been developed to identify and confirm *Salmonella* [29, 39, 98] quickly.

DNA hybridization and PCR are two more methods that may be used to identify *S. enterica* [116, 117]. Amplification and analysis of strain variation may be accomplished by using gene-specific primers in PCR testing. It can improve the detection and characterization of pathogenic bacteria by targeting species-specific DNA regions and specific pathogenicity traits, such as genes that code for toxins, virulence factors, or major antigens [84, 118, 119]. This makes it possible for it to improve the detection of pathogenic bacteria. Other *Salmonella* strain typing methods include utilizing antibiotic resistance genes as epidemiological markers using multilocus sequence typing [5, 7, 111].

These methods examine the DNA sequences of a series of housekeeping, ribosomal, and virulent genes, and therefore making isolates distinction based on the molecular analysis. This uses short sequence repeat motifs as a target to type isolates.

The polymerase chain reaction (PCR) and real-time PCR are being investigated as potential diagnostic tools for enteric fever [112]. In theory, nucleic acid amplification tests (NAATs) might amplify DNA from bacteria that are either dead or incapable of being cultured, hence rectifying low culture positives caused by antibiotic pretreatment [113]. According to research [114], the test sensitivity limitations for a PCR technique are the same as those for a culture approach. Culture and PCR are combined in some methodologies. The adoption of NAATs in developing countries is expected to be hampered because of the high cost and lack of laboratory infrastructure [120]. The effectiveness of NAATs for the diagnosis of enteric fever has been the subject of some research. The flagellin genes (*fliC-d* for *S. Typhi* and *fliC-a* for *S. Paratyphi* A) are most often targeted by PCR [121, 122]. In a study of blood PCR testing for enteric fever, researchers found that although all tests were 100 percent specific, their sensitivities differed [83, 123]. The sensitivity is considerable in most studies to be more than 90% [20] in persons with positive blood culture, but it is lower (3–13%) in those without clinical symptoms [14].

Additionally, PCR tests focusing on fliC have been applied to urine, and the findings have been favorable [124]. The primary benefit of PCR over other identification methods, like culture and conventional methods, is that it produces findings much more quickly [124]. PCR requires specialist laboratory equipment, which might be difficult in regions where typhoid fever is prevalent [14, 39]. Whole-genome sequencing (WGS) has revolutionized how antimicrobial resistance is studied [125, 126]. It has enabled the detection of resistance genes even before they are expressed and has also played a very important role in epidemiological studies of antimicrobial resistance *Salmonella* [127]. These techniques have helped develop newer diagnostics and are being explored in vaccine candidate development for several *Salmonella* species other than Typhi [128].

Both phenotypic and genotypic methods detect resistance genes or resistance mechanisms in bacteria, specifically in *S. enterica* [17]. The phenotypic method explores this bacteria's expression of certain traits to detect a resistance mechanism [122]. For example, the resistance by *Salmonella* to third-generation cephalosporin is an indication of possible possession of extended-spectrum beta-lactamases (ESBLs) [129]. *Salmonella* resistance to Meropenem is an indication to been carbapenemase-producing [130]. Genotypic techniques are employed to confirm the phenotypic detection of an antibiotic-resistance mechanism [131]. Some of these genotypic methods for antibiotic resistance genes include whole genome sequencing and polymerase chain reaction application. Whole genome sequencing is expensive, and as such, it is not routinely used to detect resistance genes. Polymerase chain reaction (PCR) is the method of choice for laboratories with the capacity [132]. Quantitative and conventional PCR is used. Quantitative PCR uses specific primers and probes to detect the resistance or virulence gene of interest. Conventional PCR, which is mostly available and more cost-effective, uses specific primers, and the products are visualized in a gel documentation system after amplification [133].

2.6 Prevention and control of resistant antibiotics and virulent *S. enterica* across one-health sectors

The prevention of *S. enterica* infection involves proper co-ordination of preventive measures across the humans and their activities, that is, agriculture, animal rearing for food, the use of animals as companions or pets, and management of environment sector to detect and eliminate any threat [100, 134] of *Salmonella* [100, 134]. The different levels of prevention of *S. enterica* pathogens include: prevention of *S. enterica* pathogens from farms to human via food or through the contaminated environment via poor waste management [120, 135]. The proper management of farms to eliminate pathogenic bacteria from the animal facilities is one effective way of managing *Salmonella* outbreaks. Also, prosper handling of food processing and animal products contributed to reduced outbreaks [10, 136]. Some keys *Salmonella* infection outbreaks were associated with transmission from processed food animals like chickens, pork, and other meat products. The use of animal wastes as fertilizers has also been associated with *S. enterica* outbreaks [76]. Full implementation of good hygiene practices across all sectors, including house hygiene practices and deployment of WASH in all sectors, will eliminate *S. enterica* from the food chain and all possible transmission avenues.

By consuming contaminated food or drink, enteric fever is most often spread from person to person [134]. In the past, enteric fever was common in Western Europe and the United States [75]. Despite this, pasteurization of milk and other dairy products, the removal of human feces in the food-manufacturing process, and good food and water cleanliness have all contributed to a considerable decline in the prevalence of *Salmonella* infection [137]. There was a decrease in the number of *Salmonella* illnesses reported in Latin America simultaneously as sanitary techniques were implemented [137]. Giving access to clean water and food, maintaining proper sanitation, and administering typhoid vaccinations are the best ways to avoid enteric fever. Making and ensuring that water meant for human consumption is safe is the main goal of eliminating possible vectors for the transmission of typhoid *Salmonella* and non-typhoid *Salmonella* (NTS). This important objective has been easily achieved in wealthy nations like Europe and the United States but not in developing or underdeveloped countries [138]. In addition to water, a variety of foods may include *Salmonella* species. However, they are often found in poultry, eggs, and dairy products [139]. The adoption of proper food handling and cooking practices has been proposed to prevent bacterial contamination of food. Due to its efficacy in reducing the risk of food contamination, food irradiation has attracted considerable interest and support in several countries. Several public health agencies, including the WHO and the CDC, have approved irradiating food. Still, due to the risk posed by radioactivity, it is only partially used in certain parts of Europe and the United States [103]. Vaccination is one of the best methods to protect against enteric fever [140]. The inactive parenteral and oral live attenuated vaccines are the two immunization types that may presently be utilized to prevent enteric fever. However, these authorized immunizations are exclusively used in infants and are ineffective at preventing diseases by *S. enterica* subspecies Paratyphi and NTS. Limiting the erroneous use of antibiotics in food animals and the feed they ingest is one approach that is good for NTS.

Hazard analysis and critical control points (HACCP) are advantageous since it is an efficient strategy for minimizing risk and maximizing product security [141]. The HACCP is employed at various stages of the One-Health sector. This is important to avoid cross-contamination or transfer of pathogenic *S. enterica* from the environment to processed food, humans, and vice versa. One way of implementing HACCP in the animal sector is to ensure that *Salmonella* pathogens are not released into the environment [142]. And everyone involved in the processing steps of food are tested for *S. enterica* to avoid shedding in processed food [143]. Implementing HACCP has several advantages, including eliminating prejudice and providing a framework for prioritizing choices. HACCP helps ensure that only those with the necessary knowledge, skills, and experience are responsible for food safety. With HACCP in place, there is concrete proof of your food safety management, which will be useful in court in the case of any litigation. After the initial investment in implementing HACCP, the system may be very cost-efficient. As a result of HACCP, food manufacturers may fulfill their mandated duty to create healthy, wholesome fares in compliance with applicable regulations [144]. Applying HACCP's procedures and guidelines almost guarantees better results every time. This is mostly attributable to people's heightened sensitivity to risks and the fact that they come from all walks of the operation. The HACCP principles and the requisite support mechanisms for a robust food safety program form the basis of the Global Food Safety Initiative (GFSI) by ensuring the absence of *Salmonella* pathogens in processed food, poultry farms, and the food chain [142, 145].

3. Conclusion

The distribution of *Salmonella* subspecies capable of causing infections has been found in humans, animals, and environments. *Salmonella* genes such as *sipA*, *sipB*, *sipC*, *sifA*, *hilA*, *hilC*, *hilD*, as well as *invA*, *spiC*, and *invF* have been linked to epidemiologic of virulent *S. enterica*. Some of the *Salmonella* from all these sources tested positive for the beta-lactamase *TEM* enzyme. To detect *S. enterica*, *invA* has been found valuable in detecting *S. enterica* contamination in food products and the environment. The *invA* gene has been made into devices and diagnostics for diagnosing infections in the bloodstream, environmental contamination, and water-processing plants. These factors have also been utilized in investigating outbreaks and infection tracing and tracking, especially in food-processing industries. The detection of this genes even without viable growth of *S. enterica* has helped control and contain outbreaks. Genes for resistance to fluoroquinolones mediated by plasmids has also been widely found in *Salmonella* species across the One-Health sector. According to research findings, most *Salmonellae* are obtained from animals, and the environment carries the virulence genes essential to induce infections in humans. Extensively drug-resistant (XDR) *Salmonella typhi* is now a serious problem in some countries, multidrug-resistant (MDR) has grown in prevalence, and *S. enterica* has evolved resistance to an increasing number of antibiotic classes. Extensively drug-resistant (XDR) *S. typhi*, so designated due to its exhibited resistance to the recommended drugs for typhoid fever, including third-generation cephalosporin, has become a serious issue that highlights the urgency in deploying the Vi-conjugate vaccines.

Conflict of interest

The authors declare no conflict of interest.

Author details

Leonard I. Uzairue[1,2*] and Olufunke B. Shittu[1]

1 Department of Microbiology, Federal University of Agriculture, Abeokuta, Ogun State, Nigeria

2 Department of Medical Laboratory Sciences, Federal University Oye Ekiti, Ekiti State, Nigeria

*Address all correspondence to: leonard.uzairue@fuoye.edu.ng; uzairue.leonard@gmail.com

IntechOpen

References

[1] Kiran Y, Yadav SK, Geeta P. A comparative study of typhidot and widal test for rapid diagnosis of typhoid fever. International Journal of Current Microbiology and Applied Sciences. 2015;**4**:34-38

[2] WHO. *Salmonella* (non-typhoidal). Vol. 1. Geneva: WHO; 2017

[3] Trinh P, Zaneveld JR, Safranek S, et al. One health relationships between Human, animal, and environmental microbiomes: A mini-review. Frontiers in Public Health. 2018;**6**:1-9

[4] Fadlallah SM, Shehab M, Cheaito K, et al. Molecular epidemiology and antimicrobial resistance of *Salmonella* species from clinical specimens and food items in Lebanon. Journal of Infection in Developing Countries. 2017;**11**:19-27. DOI: 10.3855/jidc.7786

[5] Kristinsson KG, Georgsson F. Infection risks associated with importation of fresh food in Iceland. Læknablaðið. 2015;**101**:313-319

[6] Feasey NA, Masesa C, Jassi C, et al. Three epidemics of invasive multidrug-resistant *Salmonella* bloodstream infection in Blantyre, Malawi, 1998-2014. Clinical Infectious Diseases. 2015;**61**:S363-S371

[7] Andrews JR, Ryan ET. Diagnostics for invasive *Salmonella* infections: Current challenges and future directions. (special issue: Global progress on use of vaccines for invasive *Salmonella* infections.). Vaccine. 2015;**33**:C8-C15

[8] Irwin Alec, Berthe Franck Cesar Jean, Le Gall Francois G., Marquez PV. Drug-resistant infections: A threat to our economic future: Executive summary (English). 2017;**2**:1-17

[9] Nógrády N, Imre A, Kostyák Á, et al. Molecular and pathogenic characterization of *Salmonella enterica* Serovar Bovismorbificans strains of animal, environmental, food, and Human origin in Hungary. Foodborne Pathogens and Disease. 2010;**7**:507-513

[10] Banerjee S, Ooi MC, Shariff M, et al. Antibiotic resistant *Salmonella* and *Vibrio* associated with farmed *Litopenaeus vannamei*. Scientific World Journal. 2012;**2012**:130136

[11] Gharieb RM, Tartor YH, Khedr MHE. Non-Typhoidal *Salmonella* in poultry meat and diarrhoeic patients: Prevalence, antibiogram, virulotyping, molecular detection and sequencing of class I integrons in multidrug resistant strains. Gut Pathogens. 2015;**7**:34. DOI: 10.1186/s13099-015-0081-1

[12] Chen W, Fang T, Zhou X, et al. IncHI2 plasmids are predominant in antibiotic-resistant *Salmonella* isolates. Frontiers in Microbiology. 2016;**7**:1566. DOI: 10.3389/fmicb.2016.01566

[13] Obaro SK, Hassan-Hanga F, Olateju EK, et al. *Salmonella bacteremia* among children in central and Northwest Nigeria, 2008-2015. Clinical Infectious Diseases. 2015;**61**:S325-S331

[14] Crump JA, Sjölund-Karlsson M, Gordon MA, et al. Epidemiology, clinical presentation, laboratory diagnosis, antimicrobial resistance, and antimicrobial management of invasive *Salmonella* infections. Clinical Microbiology Reviews. 2015;**28**:901-937

[15] Deng Y, Bao X, Ji L, et al. Resistance integrons: Class 1, 2 and 3 integrons. Annals of Clinical Microbiology and Antimicrobials. 2015;**14**:45. DOI: 10.1186/s12941-015-0100-6

[16] Wellington EMH, Boxall ABA, Cross P, et al. The role of the natural environment in the emergence of antibiotic resistance in gram-negative bacteria. The Lancet Infectious Diseases. 2013;**13**:155-165

[17] Ramachandran A, Shanthi M, Sekar U. Detection of blaCTX-M extended spectrum betalactamase producing *Salmonella enterica* serotype typhi in a tertiary care Centre. Journal of Clinical Diagnostic Research. 2017;**11**:DC21-DC24

[18] Salisbury AM, Bronowski C, Wigley P. *Salmonella virchow* isolates from human and avian origins in England: Molecular characterization and infection of epithelial cells and poultry. Journal of Applied Microbiology. 2011;**111**:1505-1514

[19] Matias CAR, Pereira IA, De Araújo MDS, et al. Characteristics of *Salmonella* spp. isolated from wild birds confiscated in illegal trade markets, Rio de Janeiro, Brazil. BioMed Research International. 2016;**2016**:3416864. DOI: 10.1155/2016/3416864

[20] Kariuki S, Gordon MA, Feasey N, et al. Antimicrobial resistance and management of invasive *Salmonella* disease. Vaccine. 2015;**33**:S21-S29

[21] Bailey A, Scott B. Diagnostic Microbiology. Eleventh ed. London, UK: Oxford University Press (OUP); 2002

[22] Geraldine MM, Raúl RA, Ana CO, et al. Identification of *Salmonella enteritidis* and *Salmonella typhimurium* in Guinea pigs by the multiplex PCR. Rev Investig Vet del Peru. 2017;**28**:411-417. DOI: 10.15381/rivep.v28i2.13074

[23] Dongol S, Thompson CN, Clare S, et al. The microbiological and clinical characteristics of invasive *Salmonella* in gallbladders from cholecystectomy patients in Kathmandu, Nepal. PLoS One. 2012;**7**:e47342. DOI: 10.1371/journal.pone.0047342

[24] Nga TVT, Karkey A, Dongol S, et al. The sensitivity of real-time PCR amplification targeting invasive *Salmonella* serovars in biological specimens. BMC Infectious Diseases. 2010;**10**:125. DOI: 10.1186/1471-2334-10-125

[25] Santos RL, Mikoleit ML, Unit S, et al. Characterization of *Salmonella* isolates from retail foods based on serotyping, pulse field gel electrophoresis, antibiotic resistance and other phenotypic properties. Applied and Environmental Microbiology. 2014;**77**:187-219

[26] Venkatesan N, Krishnakumar S, Deepa PR, et al. Molecular deregulation induced by silencing of the high mobility group protein A2 gene in retinoblastoma cells. Molecular Vision. 2012;**18**:2420-2437

[27] Berhane A, Russom M, Bahta I, et al. Rapid diagnostic tests failing to detect plasmodium falciparum infections in Eritrea: An investigation of reported false negative RDT results. Malaria Journal. 2017;**16**:105. DOI: 10.1186/s12936-017-1752-9

[28] Boyd DA, Shi X, Hu QH, et al. *Salmonella* genomic island 1 (SGI1), variant SGI1-I, and new variant SGI1-O in Proteus mirabilis clinical and food isolates from China. Antimicrobial Agents and Chemotherapy. 2008;**52**:340-344

[29] Felgner J, Jain A, Nakajima R, et al. Development of ELISAs for diagnosis of acute typhoid fever in Nigerian children. PLoS Neglected Tropical Diseases. 2017;**11**:e0005679. DOI: 10.1371/journal.pntd.0005679

[30] Andrews JR, Ryan ET. Diagnostics for invasive *Salmonella* infections: Current challenges and future directions. Vaccine. 2015;**33**:C8-C15

[31] Li Q, Hu Y, Chen J, et al. Identification of *Salmonella enterica*

serovar Pullorum antigenic determinants expressed in vivo. Infection and Immunity. 2013;**81**:119-127. DOI: 10.1128/IAI.00145-13

[32] Wilson RL, Elthon J, Clegg S, et al. *Salmonella enterica* serovars gallinarum and pullorum expressing *Salmonella enterica* serovar typhimurium type 1 fimbriae exhibit increased invasiveness for mammalian cells. Infection and Immunity. 2000;**68**:4782-4785. DOI: 10.1128/IAI.68.8.4782-4785.2000

[33] Wigley P, Berchieri AJ, Page KL, et al. *Salmonella enterica* serovar pullorum persists in splenic macrophages and in the reproductive tract during persistent, disease-free carriage in chickens. Infection and Immunity. 2001;**69**:7873-7879. DOI: 10.1128/IAI.69.12.7873-7879.2001

[34] Rodriguez J, Nonaka D, Kuhn E, et al. Combined high-grade basal cell carcinoma and malignant melanoma of the skin ("malignant basomelanocytic tumor"): Report of two cases and review of the literature. The American Journal of Dermatopathology. 2005;**27**:314-318

[35] Tennant SM, Diallo S, Levy H, et al. Identification by PCR of non-typhoidal *Salmonella enterica* serovars associated with invasive infections among febrile patients in Mali. PLoS Neglected Tropical Diseases. 2010;**4**:e621. DOI: 10.1371/journal.pntd.0000621

[36] Suez J, Porwollik S, Dagan A, et al. Virulence gene profiling and pathogenicity characterization of non-Typhoidal *Salmonella* accounted for invasive disease in humans. PLoS One. 2013;**8**:e58449. DOI: 10.1371/journal.pone.0058449

[37] Tennant SM, MacLennan CA, Simon R, et al. Nontyphoidal *Salmonella* disease: Current status of vaccine research and development. Vaccine. 2016;**34**:2907-2910. DOI: 10.1016/j.vaccine.2016.03.072

[38] Sangal V, Harbottle H, Mazzoni CJ, et al. Evolution and population structure of *Salmonella enterica* serovar Newport. Journal of Bacteriology. 2010;**192**:6465-6476. DOI: 10.1128/JB.00969-10

[39] Okoro CK, Kingsley RA, Connor TR, et al. Intracontinental spread of human invasive *Salmonella typhimurium* pathovariants in sub-Saharan Africa. Nature Genetics. 2012;**44**:1215-1221

[40] Feasey NA, Dougan G, Kingsley RA, et al. Invasive non-typhoidal *Salmonella* disease: An emerging and neglected tropical disease in Africa. Lancet. 2012;**379**:2489-2499

[41] Marks F, von Kalckreuth V, Aaby P, et al. Incidence of invasive *Salmonella* disease in sub-Saharan Africa: A multicentre population-based surveillance study. Lancet Glob Heal. 2017;**5**:e310-e323

[42] Ben Hassena A, Barkallah M, Fendri I, et al. Real time PCR gene profiling and detection of *Salmonella* using a novel target: The siiA gene. Journal of Microbiological Methods. 2015;**109**:9-15. DOI: 10.1016/j.mimet.2014.11.018

[43] Rao S, Schieber AMP, O'Connor CP, et al. Pathogen-mediated inhibition of anorexia promotes host survival and transmission. Cell. 2017;**168**:503-516. DOI: 10.1016/j.cell.2017.01.006

[44] Shippy DC, Eakley NM, Bochsler PN, et al. Biological and virulence characteristics of *Salmonella enterica* serovar Typhimurium following deletion of glucose-inhibited division (gidA) gene. Microbial Pathogenesis. 2011;**50**:303-313

[45] Fazl AA, Salehi TZ, Jamshidian M, et al. Molecular detection of invA, ssaP, sseC and pipB genes in *Salmonella typhimurium* isolated from human and poultry in Iran. African Journal of Microbiological Research. 2013;**7**:1104-1108

[46] McWhorter AR, Chousalkar KK. Comparative phenotypic and genotypic virulence of *Salmonella* strains isolated from Australian layer farms. Frontiers in Microbiology. 2015;**6**:1-14

[47] Velge P, Wiedemann A, Rosselin M, et al. Multiplicity of *Salmonella* entry mechanisms, a new paradigm for *Salmonella* pathogenesis. Microbiology. 2012;**1**:243-258. DOI: 10.1002/mbo3.28

[48] Antonio Ibarra J, Knodler LA, Sturdevant DE, et al. Induction of *Salmonella* pathogenicity island 1 under different growth conditions can affect *Salmonella*-host cell interactions in vitro. Microbiology. 2010;**156**:1120-1133

[49] Khoo CH, Cheah YK, Lee LH, et al. Virulotyping of *Salmonella enterica* subsp. enterica isolated from indigenous vegetables and poultry meat in Malaysia using multiplex-PCR. Antonie van Leeuwenhoek. International Journal of Genetics & Molecular Microbiology. 2009;**96**:441-457

[50] Lawley TD, Bouley DM, Hoy YE, et al. Host transmission of *Salmonella enterica* serovar Typhimurium is controlled by virulence factors and indigenous intestinal microbiota. Infection and Immunity. 2008;**76**:403-416. DOI: 10.1128/IAI.01189-07

[51] Hensel M. Evolution of pathogenicity islands of *Salmonella enterica*. International Journal of Medical Microbiology. 2004;**294**:95-102. DOI: 10.1016/j.ijmm.2004.06.025

[52] Izumiya H, Kuroda M, Tamamura Y, et al. Phylogenetic characterization of *Salmonella enterica* serovar Typhimurium and its monophasic variant isolated from food animals in Japan revealed replacement of major epidemic clones in the last 4 decades. Journal of Clinical Microbiology. 2018;**56**:e01758-e01717. DOI: 10.1128/jcm.01758-17

[53] Li C, Hu D, Xue W, et al. Treatment outcome of combined continuous Venovenous hemofiltration and Hemoperfusion in acute Paraquat poisoning: A prospective controlled trial. Critical Care Medicine. 2018;**46**:100-107. DOI: 10.1097/CCM.0000000000002826

[54] Walthers D, Carroll RK, Navarre WW, et al. The response regulator SsrB activates expression of diverse *Salmonella* pathogenicity island 2 promoters and counters silencing by the nucleoid-associated protein H-NS. Molecular Microbiology. 2007;**65**:477-493. DOI: 10.1111/j.1365-2958.2007.05800.x

[55] Forest CG, Ferraro E, Sabbagh SC, et al. Intracellular survival of *Salmonella enterica* serovar Typhi in human macrophages is independent of *Salmonella* pathogenicity island (SPI)-2. Microbiology. 2010;**156**:3689-3698. DOI: 10.1099/mic.0.041624-0

[56] Haneda T, Ishii Y, Shimizu H, et al. *Salmonella* type III effector SpvC, a phosphothreonine lyase, contributes to reduction in inflammatory response during intestinal phase of infection. Cellular Microbiology. 2012;**14**:485-499. DOI: 10.1111/j.1462-5822.2011.01733.x

[57] Kaur J, Jain SK. Role of antigens and virulence factors of *Salmonella enterica* serovar Typhi in its pathogenesis. Microbiological Research. 2012;**167**:199-210. DOI: 10.1016/j.micres.2011.08.001

[58] Huehn S, La Ragione RM, Anjum M, et al. Virulotyping and antimicrobial resistance typing of *Salmonella enterica*

serovars relevant to human health in Europe. Foodborne Pathogens and Disease. 2010;**7**:523-535

[59] Cao J, Xu L, Yuan M, et al. TaqMan probe real-time PCR detection of foodborne *Salmonella enterica* and its six Serovars. International Journal of Current Microbiology and Applied Sciences. 2013;**2**:1-12

[60] Ginocchio CC, Rahn K, Clarke RC, et al. Naturally occurring deletions in the centisome 63 pathogenicity island of environmental isolates of *Salmonella* spp. Infection and Immunity. 1997;**65**:1267-1272

[61] Malorny B, Fach P, Bunge C, Martin A, REP H, Malorny B, et al. Diagnostic real-time PCR for detection of *Salmonella* in food. Applied and Environmental Microbiology. 2004;**70**:7046-7052

[62] Ulaya WD. Determination of Virulence Factors in *Salmonella* Isolates of Human, Poultry and Dog Origin in Lusaka District. Lusaka, Zambia: Zambia School of Veterinary Medicine, Department of Paraclinical Studies; 2013;**2**(3):23-32

[63] Ghanizadeh A. Malondialdehyde, Bcl-2, superoxide dismutase and glutathione peroxidase may mediate the association of sonic hedgehog protein and oxidative stress in autism. Neurochemical Research. 2012;**37**:899-901

[64] Ball TA, Fedorka-Cray PJ, Horovitz J, et al. Molecular characterization of *Salmonella* spp. from cattle and chicken farms in Uganda. Online Journal of Public Health Informatics. 2018. DOI: 10.5210/ojphi.v10i1.8934

[65] Douard G, Praud K, Cloeckaert A, et al. The *Salmonella* genomic island 1 is specifically mobilized in trans by the IncA/C multidrug resistance plasmid family. PLoS One. 2010;**5**. DOI: 10.1371/journal.pone.0015302

[66] Huguet KT, Gonnet M, Doublet B, et al. A toxin antitoxin system promotes the maintenance of the IncA/C-mobilizable *Salmonella* Genomic Island 1. Scientific Reports. 2016;**6**. DOI: 10.1038/srep32285

[67] Shah M, Kathiiko C, Wada A, et al. Prevalence, seasonal variation, and antibiotic resistance pattern of enteric bacterial pathogens among hospitalized diarrheic children in suburban regions of Central Kenya. Tropical Medical Health. 2016. DOI: 10.1186/s41182-016-0038-1

[68] Jajere SM. A review of *Salmonella enterica* with particular focus on the pathogenicity and virulence factors, host specificity and antimicrobial resistance including multidrug resistance. Veterinary World. 2019;**12**:504

[69] Hall RM. *Salmonella* genomic islands and antibiotic resistance in *Salmonella enterica*. Future Microbiology. 2010. DOI: 10.2217/fmb.10.122

[70] Osaili TM, Abu Jamous DO, Obeidat BA, et al. Food safety knowledge among food workers in restaurants in Jordan. Food Control. 2013. DOI: 10.1016/j.foodcont.2012.09.037

[71] Centers for Disease Control and Prevention (CDC). Notes from the field: Human *Salmonella infantis* infections linked to dry dog food--United States and Canada. MMWR. Morbidity and Mortality Weekly Report. 2012;**2012**

[72] Obaro S, Lawson L, Essen U, et al. Community acquired bacteremia in young children from Central Nigeria--a pilot study. BMC Infectious Diseases. 2011;**11**:137

[73] Kasturi KN, Drgon T. Real-time PCR method for detection of *Salmonella*

spp. in environmental samples. Applied and Environmental Microbiology. 2017;**83**:1-12

[74] Miller T, Brockmann S, Spackova M, et al. Recurring outbreaks caused by the same *Salmonella infantis* clone in a German rehabilitation oncology clinic over at least 2002 to 2009. The Journal of Hospital Infection. 2018. DOI: 10.1016/j.jhin.2018.03.035

[75] Basler C, Forshey TM, Machesky K, et al. Notes from the field: Multistate outbreak of human *Salmonella* infections linked to live poultry from a mail-order hatchery in Ohio--February-October 2014. MMWR. Morbidity and Mortality Weekly Report. 2015;**32**:20-28

[76] Shittu OB, Uzairue LI, Ojo OE, et al. Antimicrobial resistance and virulence genes in *Salmonella enterica* serovars isolated from droppings of layer chicken in two farms in Nigeria. Journal of Applied Microbiology. 2022;**132**:3891-3906

[77] Stanaway JD, Parisi A, Sarkar K, et al. The global burden of non-typhoidal *Salmonella* invasive disease: A systematic analysis for the global burden of disease study 2017. The Lancet Infectious Diseases. 2019;**19**:1312-1324

[78] Gopinath S, Carden S, Monack D. Shedding light on *Salmonella* carriers. Trends in Microbiology. 2012. DOI: 10.1016/j.tim.2012.04.004

[79] Berger CN, Sodha SV, Shaw RK, et al. Fresh fruit and vegetables as vehicles for the transmission of human pathogens. Environmental Microbiology. 2010. DOI: 10.1111/j.1462-2920.2010.02297.x

[80] Rawat D, Nair D. Extended-spectrum ß-lactamases in gram negative bacteria. Journal of Global Infectious Diseases. 2010. DOI: 10.4103/0974-777x.68531

[81] Hawkey PM, Jones AM. The changing epidemiology of resistance. The Journal of Antimicrobial Chemotherapy. 2009. DOI: 10.1093/jac/dkp256

[82] Pal C, Bengtsson-Palme J, Kristiansson E, et al. Co-occurrence of resistance genes to antibiotics, biocides and metals reveals novel insights into their co-selection potential. BMC Genomics. 2015. DOI: 10.1186/s12864-015-2153-5

[83] Kariuki S, Gordon MA, Feasey N, et al. Antimicrobial resistance and management of invasive *Salmonella* disease HHS public access. Vaccine. 2015;**19**:21-29

[84] Vo ATT, van Duijkeren E, Fluit AC, et al. A novel *Salmonella* genomic island 1 and rare integron types in *Salmonella typhimurium* isolates from horses in the Netherlands. The Journal of Antimicrobial Chemotherapy. 2007;**59**:594-599

[85] Adesiji YO, Deekshit VK, Karunasagar I. Antimicrobial-resistant genes associated with *Salmonella* spp. isolated from human, poultry, and seafood sources. Food Science & Nutrition. 2014. DOI: 10.1002/fsn3.119

[86] Flórez AB, Alegría Á, Rossi F, et al. Molecular identification and quantification of tetracycline and erythromycin resistance genes in Spanish and Italian retail cheeses. BioMed Research International. 2014;**2014**:1-10

[87] Guerra B, Soto S, Helmuth R, et al. Characterization of a self-transferable plasmid from *Salmonella enterica* serotype typhimurium clinical isolates carrying two integron-borne gene cassettes together with virulence and drug resistance genes. Antimicrobial Agents and Chemotherapy. 2002. DOI: 10.1128/AAC.46.9.2977-2981.2002

[88] Tran-Dien A, Le Hello S, Bouchier C, et al. Early transmissible ampicillin resistance in zoonotic *Salmonella enterica* serotype Typhimurium in the late 1950s: A retrospective, whole-genome sequencing study. The Lancet Infectious Diseases. 2018. DOI: 10.1016/S1473-3099(17)30705-3

[89] Abakpa GO, Umoh VJ, Ameh JB, et al. Diversity and antimicrobial resistance of *Salmonella enterica* isolated from fresh produce and environmental samples. Environmental Nanotechnology, Monitor Management. 2015;**3**:38-46

[90] Marin C, Martín-Maldonado B, Cerdà-Cuéllar M, et al. Antimicrobial resistant *Salmonella* in chelonians: Assessing its potential risk in zoological institutions in Spain. Veterinary Science. 2022;**9**. DOI: 10.3390/VETSCI9060264

[91] Simpson KMJ, Mor SM, Ward MP, et al. Genomic characterisation of *Salmonella enterica* serovar Wangata isolates obtained from different sources reveals low genomic diversity. PLoS One. 2020;**15**:e0229697

[92] Hermans APHM, Beuling AM, van Hoek AHAM, et al. Distribution of prophages and SGI-1 antibiotic-resistance genes among different *Salmonella enterica* serovar Typhimurium isolates. Microbiology. 2006;**152**:2137-2147

[93] de Jong HK, Parry CM, van der Poll T, et al. Host-pathogen interaction in invasive salmonellosis. PLoS Pathogens. 2012. DOI: 10.1371/journal.ppat.1002933

[94] Harish B, Menezes G. Antimicrobial resistance in typhoidal Salmonellae. Indian Journal of Medical Microbiology. 2011. DOI: 10.4103/0255-0857.83904

[95] Akinyemi KO, Oyefolu AOB, Mutiu WB, et al. Typhoid fever: Tracking the trend in Nigeria. The American Journal of Tropical Medicine and Hygiene. 2018. DOI: 10.4269/ajtmh.18-0045

[96] Kasper MR, Sokhal B, Blair PJ, et al. Emergence of multidrug-resistant *Salmonella enterica* serovar Typhi with reduced susceptibility to fluoroquinolones in Cambodia. Diagnostic Microbiology and Infectious Disease. 2010;**66**:207-209

[97] Britto CD, Wong VK, Dougan G, et al. A systematic review of antimicrobial resistance in *Salmonella enterica* serovar Typhi, the etiological agent of typhoid. PLoS Neglected Tropical Diseases. 2018;**12**:e0006779

[98] Mastrorilli E, Petrin S, Orsini M, et al. Comparative genomic analysis reveals high intra-serovar plasticity within *Salmonella napoli* isolated in 2005-2017. BMC Genomics. 2020;**21**:1-16

[99] Kariuki S, Onsare RS. Epidemiology and genomics of invasive nontyphoidal *Salmonella* infections in Kenya. Clinical Infectious Diseases. 2015;**61**:S317-S324

[100] Marin C, Lorenzo-Rebenaque L, Laso O, et al. Pet reptiles: A potential source of transmission of multidrug resistant *Salmonella*. Frontiers in Veterinary Science. 2021;7:1157

[101] Zorgani A, Ziglam H. Typhoid fever: Misuse of Widal test in Libya. Journal of Infection in Developing Countries. 2014;**8**:680-687

[102] Castonguay-Vanier J, Davong V, Bouthasavong L, et al. Evaluation of a simple blood culture amplification and antigen detection method for diagnosis of *Salmonella enterica* serovar Typhi bacteremia. Journal of Clinical Microbiology. 2013;**51**:142-148

[103] MacFadden DR, Bogoch II, Andrews JR. Advances in diagnosis,

treatment, and prevention of invasive *Salmonella* infections. Current Opinion in Infectious Diseases. 2016;**29**:453-458

[104] Odeyemi OA. Bacteriological safety of packaged drinking water sold in Nigeria: Public health implications. Springerplus. 2015. DOI: 10.1186/s40064-015-1447-z

[105] Gunn JS, Marshall JM, Baker S, et al. *Salmonella* chronic carriage: Epidemiology, diagnosis, and gallbladder persistence. Trends in Microbiology. 2014;**22**:648-655

[106] Saporito L, Colomba C, Titone L. Typhoid fever. In: International Encyclopedia of Public Health. London, UK: Oxford Press; 2016

[107] Scott S. Bailey and Scott's Diagnostic Microbiology. Amsterdam, The Netherlands: Elsevier; 2014

[108] Malehmir S, Ranjbar R, Harzandi N. The molecular study of antibiotic resistance to quinolones in *Salmonella enterica* strains isolated in Tehran, Iran. Open Microbiology Journal. 2017;**11**:189-194

[109] Clinical Laboratory Standard Institue (CLSI). Performance Standards for Antimicrobial Susceptibility Testing. 26th ed. CLSI Supplement M100S. Wayne, USA: CLSI; 2018

[110] Chen S, Zhao S, White DG, et al. Characterization of *Salmonella* serovars isolated from retail meats. Applied and Environmental Microbiology. 2004;**70**:1-7

[111] Hur J, Jawale C, Lee JH. Antimicrobial resistance of *Salmonella* isolated from food animals: A review. Food Research International. 2012;**45**:819-830

[112] Kaprou GD, Papadakis G, Papageorgiou DP, et al. Miniaturized devices for isothermal DNA amplification addressing DNA diagnostics. Microsystem Technologies. 2016;**22**:1529-1534

[113] Mayboroda O, Katakis I, O'Sullivan CK. Multiplexed isothermal nucleic acid amplification. Analytical Biochemistry. 2018;**545**:20-30

[114] Tanner NA, Zhang Y, Evans TC. Visual detection of isothermal nucleic acid amplification using pH-sensitive dyes. BioTechniques. 2015;**58**:59-68

[115] CDC. Serotypes and the importance of serotyping *Salmonella*, *Salmonella* atlas, reports and publications, *Salmonella*, CDC. Centers for Disease Control Prevention. 2015. DOI: 10.1208/s12249-010-9573-y

[116] Mercer R, Nguyen O, Ou Q, et al. Functional analysis of genes encoded by the locus of heat resistance in Escherichia coli. Applied and Environmental Microbiology. 2017. DOI: 10.1128/AEM.01400-17

[117] Ammar AM, Mohamed AA, El-Hamid MIA, et al. Virulence genotypes of clinical *Salmonella* serovars from broilers in Egypt. Journal of Infection in Developing Countries. 2016. DOI: 10.3855/jidc.7437

[118] Cajetan ICI, Bassey BE, Florence IN, et al. Prevalence and antimicrobial susceptibility of *Salmonella* species associated with childhood acute gastroenteritis in Federal Capital Territory Abuja, Nigeria. British Microbiology Research Journal. 2013;**3**:431-439

[119] Karmi M. Detection of virulence gene (inva) in *Salmonella* isolated from meat and poultry products. International Journal of Genetics. 2013;**3**:7-12

[120] Crump JA, Heyderman RS. A perspective on invasive *Salmonella*

disease in Africa. Clinical Infectious Diseases. 2015;**61**:S235-S240

[121] Weile J, Knabbe C. Current applications and future trends of molecular diagnostics in clinical bacteriology. Analytical and Bioanalytical Chemistry. 2009;**394**:731-742

[122] Capuano F, Mancusi A, Capparelli R, et al. Characterization of drug resistance and virulotypes of *Salmonella* strains isolated from food and humans. Foodborne Pathogens and Disease. 2013;**10**:963-968

[123] Parry CM. Antimicrobial drug resistance in *Salmonella enterica*. Current Opinion in Infectious Diseases. 2003;**16**:467-472

[124] Tennant SM, Toema D, Qamar F, et al. Detection of typhoidal and paratyphoidal *Salmonella* in blood by real-time polymerase chain reaction. Clinical Infectious Diseases. 2015;**61**:S241-S250

[125] Aworh MK, Kwaga JKP, Hendriksen RS, et al. Genetic relatedness of multidrug resistant Escherichia coli isolated from humans, chickens and poultry environments. Antimicrobial Resistance and Infection Control. 2021;**10**:1-13

[126] Messens W, Hugas M, Afonso A, et al. Advancing biological hazards risk assessment. EFSA Journal. 2019;**17**. DOI: 10.2903/j.efsa.2019.e170714

[127] Bernreiter-hofer T, Schwarz L, Müller E, et al. The pheno- and genotypic characterization of porcine escherichia coli isolates. Microorganisms. 2021;**9**:1-21

[128] Crump JA, Wain J. *Salmonella*. In: International Encyclopedia of Public Health. 2016

[129] Opintan JA, Newman MJ, Arhin RE, et al. Laboratory-based nationwide surveillance of antimicrobial resistance in Ghana. Infection and Drug Resistance. 2015;**8**:379-389

[130] Algammal AM, Hetta HF, Batiha GE, et al. Virulence-determinants and antibiotic-resistance genes of MDR-E. coli isolated from secondary infections following FMD-outbreak in cattle. Scientific Reports. 2020;**10**. DOI: 10.1038/S41598-020-75914-9

[131] Ifeanyi SS. Molecular detection of some virulence genes in *Salmonella* spp. isolated from food samples in Lagos, Nigeria. Animal Veterinary Science. 2015. DOI: 10.11648/j.avs.20150301.15

[132] Zankari E, Hasman H, Cosentino S, et al. Identification of acquired antimicrobial resistance genes. The Journal of Antimicrobial Chemotherapy. 2012. DOI: 10.1093/jac/dks261

[133] Hamid N, Jain SK. Characterization of an outer membrane protein of *Salmonella enterica* serovar Typhimurium that confers protection against typhoid. Clinical and Vaccine Immunology. 2008. DOI: 10.1128/CVI.00093-08

[134] Langendorf C, Le Hello S, Moumouni A, et al. Enteric bacterial pathogens in children with diarrhea in Niger: Diversity and antimicrobial resistance. PLoS One. 2015;**10**. DOI: 10.1371/journal.pone.0120275

[135] Mandomando I, Bassat Q, Sigaúque B, et al. Invasive *Salmonella* infections among children from rural Mozambique, 2001-2014. Clinical Infectious Diseases. 2015;**61**:S339-S345

[136] Fagbamila IO, Mancin M, Barco L, et al. Investigation of potential risk factors associated with *Salmonella* presence in commercial laying hen farms in Nigeria. Preventive Veterinary

Medicine. 2018. DOI: 10.1016/j.prevetmed.2018.02.001

[137] Haselbeck AH, Panzner U, Im J, et al. Current perspectives on invasive nontyphoidal *Salmonella* disease. Current Opinion in Infectious Diseases. 2017;**30**:498-503

[138] Morpeth SC, Ramadhani HO, John AC. Invasive non-Typhi *Salmonella* disease in Africa. Clinical Infectious Diseases. 2009;**49**:606-611

[139] Pui CF, Wong WC, Chai LC, et al. *Salmonella*: A foodborne pathogen. International Food Research Journal. 2011;**10**:18

[140] Sánchez-Vargas FM, Abu-El-Haija MA, Gómez-Duarte OG. *Salmonella* infections: An update on epidemiology, management, and prevention. Travel Medicine and Infectious Disease. 2011;**9**:263-277

[141] United States Department of Agriculture. Generic HACCP Model for Poultry Slaughter. HACCP-5; Washington DC, USA: Department of Agriculture; 1999. pp. 1-35

[142] Northcutt JK, Russell SM. General guidelines for implementation of HACCP in a poultry processing plant. College of Agricultural and Environmental Journal. 2010;**3**:1-8

[143] Hamrin P, Hoeft B. Quality control throughout the production process of infant food. Annals of Nutrition & Metabolism. 2012;**60**:208-210

[144] Okonko IO, Adejoye OD, Ogun a a et al. Hazards Analysis Critical Control Points (HACCP) and Microbiology Qualities of Sea-Foods as Affected by Handler' s Hygience in Ibadan and Lagos vol. 3. Nigeria; Onward press; 2009. PP. 35-50

[145] Henry CJK, Xin JLW. Application of hazard analysis critical control point in the local manufacture of ready-to-use therapeutic foods (RUTFs). Food and Nutrition Bulletin. 2014;**35**:S57-S63

Chapter 4

Involvement of CRISPR-Cas Systems in *Salmonella* Immune Response, Genome Editing, and Pathogen Typing in Diagnosis and Surveillance

Ruimin Gao and Jasmine Rae Frost

Abstract

Clustered regularly interspaced short palindromic repeats (CRISPR) and their associated *cas* genes (CRISPR-Cas) provide acquired immunity in prokaryotes and protect microbial cells against infection by foreign organisms. CRISPR regions are found in bacterial genomes including *Salmonella* which is one of the primary causes of bacterial foodborne illness worldwide. The CRISPR array is composed of a succession duplicate sequences (repeats) which are separated by similar sized variable sequences (spacers). This chapter will first focus on the CRISPR-Cas involved in *Salmonella* immune response. With the emergence of whole genome sequencing (WGS) in recent years, more *Salmonella* genome sequences are available, and various genomic tools for CRISPR arrays identification have been developed. Second, through the analysis of 115 *Salmonella* isolates with complete genome sequences, significant diversity of spacer profiles in CRISPR arrays. Finally, some applications of CRISPR-Cas systems in *Salmonella* are illustrated, which mainly includes genome editing, CRISPR closely relating to antimicrobial resistance (AMR), CRISPR typing and subtyping as improved laboratory diagnostic tools. In summary, this chapter provides a brief review of the CRISPR-Cas system in *Salmonella*, which enhances the current knowledge of *Salmonella* genomics, and hold promise for developing new diagnostics methods in improving laboratory diagnosis and surveillance endeavors in food safety.

Keywords: *Salmonella*, CRISPR-Cas, WGS, CRISPR typing, immune response, genome editing, AMR, surveillance

1. Introduction

Clustered regularly interspaced short palindromic repeats (CRISPR) and their associated *cas* genes (CRISPR-Cas) are a family of DNA sequences, as an adaptive immune system, which protects microbial cells against infection by foreign nucleotide elements including plasmids and phages [1]. CRISPR are widespread in prokaryotes, and found in approximately 50% bacterial genomes including *Salmonella* belonging to the family of *Enterobacteriaceae* [2], which is a primary cause of bacterial foodborne illness worldwide.

Through a computational analysis of CRISPR-Cas systems, a classification system was determined based on the gene or genes encoding the effector molecules [3]. This analysis determined that CRISPR-Cas systems can fall into two classes; class 1 systems (types I, III and IV) which use a complex of multiple proteins to degrade foreign nucleic acids, and class 2 systems (types II, V, and VI) which only require a single large Cas protein (**Figure 1**) [4]. The six types of systems are further divided into 36 subtypes (**Figure 1**) [3, 5–10]. Fully functional CRISPR-Cas systems consist of CRISPR array, Cas proteins and AT-rich leader sequences. The phylogeny of CRISPR and associated *cas* genes could reflect different evolutionary histories [11, 12]. The CRISPR array is composed of a succession of highly conserved direct repeats (DR) of 24–47 bp separated by similar sized unique sequences (spacers) [13]. The *cas* genes are usually located near the CRISPR locus but can also be located elsewhere on the genomes. Cas proteins perform many functions, for instance, destroying foreign genomes, mediating foreign sequences acquisition into CRISPR array, and assisting the mature CRISPR RNAs (crRNAs) production [14–17]. CRISPR-Cas systems adapt by acquiring new spacers at the leader proximal end [1]. The units (DR+spacer) may target an invading piece of DNA and result in its degradation via a proposed mechanism similar to RNA interference. The distribution of CRISPR-Cas loci in different *Enterobacteriaceae* families showed that *Eshcerichia* and *Salmonella* are the top two genera containing type I-E subtypes [2]. CRISPR are reported in two pairs of loci in *Escherichia*, and one single pair in *Salmonella*, with each pair loci showing similar repeat sequences and putative linkage to common *cas* genes [11].

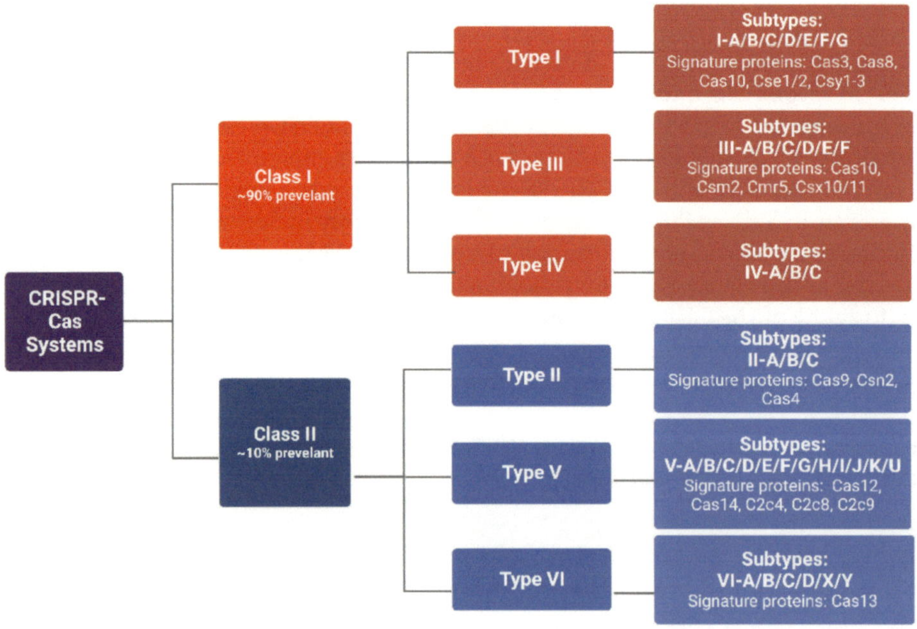

Figure 1.
General classification of CRISPR-Cas systems. Two classes—indicated by the red and blue colouring—cover six types. A total of 36 subtypes are further divided under the six types with the known signature proteins listed.

It has been shown that CRISPR spacer DNA sequences are molecular signatures used for pathogen subtyping [18] and CRISPR content correlates with the pathogenic potential of bacteria as CRISPR-Cas limits acquisition of foreign nucleotides in bacteria [19]. It has been demonstrated that there is a negative correlation between the amount of CRISPR units and pathogenicity traits, i.e. a higher number of virulence factors with lower CRIPSR repeats [19]. Based on the specific spacers, CRISPR array based quantitative PCR can be used to detect the presence of different serotypes in both *Escherichia* and *Salmonella*, with prominent sensitivity and specificity [20, 21]. Hence, the CRISPRs represent a promising genetic marker and diagnostic tool for comparative and evolutionary analysis of closely related bacterial strains [2]. Furthermore, the recognition of CRISPR-Cas9 by the Nobel Prize in Chemistry in 2020 [22] reflected its outstanding impact in genome editing field. Originated from bacteria, the CRISPR technology has already been broadly applied to fungus, yeast, insects, plants, and animals [23]. This technology has also demonstrated to functionally inactivate genes in human cell lines and cells. For instance, in 2019, CRISPR was used to treat a 34-year-old patient with sickle cell disease which is a blood genetic disorder disease [24]; and in 2020, CIRSPR-modified virus was injected into a patient's eye to treat Leber congenital amaurosis [25]. In this chapter, we will mainly focus on a foodborne pathogen *Salmonella* which is a primary cause of bacterial gastroenteritis worldwide.

Salmonella enterica is a tremendously diverse species comprising six subspecies and over 2600 serovars. *S. enterica* subsp. *enterica* accounts for the majority of clinical cases of salmonellosis and the majority of serovar diversity. Serovars of Enteritidis, Typhimurium, and Heidelberg are three main ones causing human illness. This book chapter will mainly focus on CRISPR-Cas in the immune response system of *Salmonella*, as well as its application in genome editing, pathogen typing, diagnosis and surveillance.

2. CRISPR-Cas systems of *Salmonella* in comparison with other bacteria

2.1 Immune function of CRISPR-Cas systems

In prokaryotes, bacterial CRISPR-Cas systems are unique in providing adaptive immunity against exogenous nucleotides elements, by utilizing sequence-specific RNA-guided nucleases to defend against bacteriophage infection. Bacteriophages (phages) are viruses infecting bacteria, and they are the most abundant life forms on earth. Generally, three major steps are involved in the CRISPR immune functional process: (1) new spacer acquisition—Cas proteins integrate short sequences of invading DNA into the CRISPR array; (2) CRISPR expression—CRISPR arrays are transcribed and processed to produce small crRNA; (3) CRISPR interference—crRNA along with Cas nucleases target the spacer sequence, resulting in degradation of the invader's nucleotides (DNA or RNA) [26, 27].

2.2 Characterization of CRISPR loci and *cas* genes

Like many other bacteria, the *Salmonella* genome also contains CRISPR loci. It usually contains two CRISPR loci, CRISPR1 and CRISPR2, both found on the minus strand. These two loci are separated by ~16 kb and share the same consensus direct repeat sequence (29 nt). Each CRISPR loci is fairly conserved in *Salmonella*, with the CRISPR1 locus being more conserved than CRISPR2. There are eight *cas* genes—*cas3*, *cse1*, *cse2*, *cas7*, *cas5*, *cas6*, *cas1* and *cas2*, which are located upstream of CRISPR1. Among these eight genes, *cas1* and *cas2* are universal and both are present in all CRISPR-Cas systems; *cas3* is a signature gene in the type I system; the remaining *cas* genes are type I-E dependent [28]. Furthermore, the *cse2*, *cas5*, *cas6e*, *cas1*, *cas2* and *cas3* genes are crucial for the expression of a master porin regulator named OmpR which is a two-component system regulator inducing the synthesis of OmpC, MmpF, and OmpS2 portins [29]. By taking advantages of whole genome sequencing (WGS), in 2014, researchers have demonstrated two distinct *cas* gene profiles and a high diversity of length for both CRISPR arrays, among the analysis of 64 *Salmonella* serovars [30].

2.3 CRISPR and anti-CRISPR

To combat bacterial CRISPR-Cas system, numerous phages are well known to produce proteins which can block the function of CRISPR-Cas systems, i.e. anti-CRISPR function [31]. For class 1 CRISPR system, the first discovered phage-encoded anti-CRISPR protein (Acr) was from type I-F and I-E CRISPR-Cas systems in *Pseudomonas aeruginosa*; these anti-proteins encode distinct, small proteins (50–150 aa) with different sequences and structures [32, 33]. Furthermore, these anti-CRISPRs are produced from prophages (phage sequences that have integrated into bacterial genomes) and inactivate the host (bacterial) CRISPR-Cas systems [32]. For class 2 CRISPR, four unique type II CRISPR-Cas9 inhibitor proteins have been discovered from the prophage sequences integrated into another foodborne pathogen *Listeria monocytogenes* genomes, which have type II-A CRISPR-Cas systems and their spacers have been identified by various virulent, temperate phages [34, 35]. Given more than half of *L. monocytogenes* strains with cas9 contain at least one prophage-encoded inhibitor, this suggests the possibility of widespread CRIPSR-Cas9 inactivation. Two of the discovered inhibitors in *L. monocytogenes* are also able to block the *Streptococcus pyogenes* Cas9 when analyzed in *Escherichia coli* and human cells. Similarly, in *Streptococcus thermophiles*, AcrIIA6 acts as an allosteric inhibitor and induced Cas9 dimerization [36]. Thus, the concept of natural Cas9-specific "anti-CRISPRs" presents a tool which can be used to regulate the genome engineering activities of CRISPR-Cas9 [31]. Similar to *L. monocytogenes*, in different *Salmonella* serovars, they all contain various types of prophage sequences [37]. To date there is no reported anti-CRISPR proteins in *Salmonella*, though this could change as more studies are carried out.

3. Identification and characterization of CRISPR arrays

Next generation sequencing (NGS) and especially WGS has emerged in recent years and has made it possible to sequence bacterial genomes within hours, a notable accomplishment that is revolutionizing the field of microbiology [38]. With the advent of microbial WGS, new light is shed on the nature of pathogens, for instance CRISPRs, and our understanding of the biology of *Salmonella* is steadily increasing as *Salmonella* genomes are generated at a rapid rate and are deposited in public database such as National Center for Biotechnology Information (NCBI). Based on the availability of genome sequences, various genomic and bioinformatics tools have been developed for identifying the potential CRISPR arrays in *Salmonella* genomes.

3.1 *In silico* genomics based CRISPRs identification tools

An example of how the field of CRISPRs has evolved can be seen in the work done with *in silico* analysis. *In silico* identification and analyses of CRISPRs started in 1995 [39], and several CRISPR software tools have been developed since then. In April 2007, the first specific stand-alone developed tool was CRISPRFinder, which was a web tool in identifying CRISPRs [40]. CRISPRFinder was able to define DRs and extract spacers; to get the flanking sequences and to determine the leader sequences; and then BLAST the identified spacers to check if the identified DR was present in other genomes [40]. Two months later in June 2007, in order to dissect and understand CRISPR structure and flanking sequences evolution, the same group created a public database named CRISPRdb, for which CRISPRFinder was used to analyze all the available prokaryotic genomes [41]. In the same month June 2007, a tool named CRISPR Recognition Tool (CRT) for automatic detection of the CRISPR arrays was also released [42]. CRT was demonstrated to be very reliable, with significant improvements in regards to performance in measures of precision, recall and quality, as compared to the previous existed detection tools Patscan and Pilercr [42]. In April 2008, a website based tool CRISPRcompar was created to compare CRISPRs that present a useful genetic marker for comparative analysis of closely related bacterial strains; this facilitated the development of CRISPR based pathogen typing processes [43]. More CRISPR-Cas related online tools can be found in CRISPR-Cas++, which are available at https://crisprcas.i2bc.paris-saclay.fr/. In 2018, an improved CRISPRs identification tool CRISPRCasViewer was released, which can predict CRISPR orientation, possess the latest classification scheme, and facilitate expert validation based on a rating system [44]. Alternatively, the public available "standalone" Unix/Linux version of CRISPRCasViewer can also be downloaded and installed in high-performance computing cluster bioinformatics infrastructures (https://github.com/dcouvin/CRISPRCasViewer). Thus, with the availability of all the genomic tools, the CRISPRs and *cas* genes present in each *Salmonella* isolate are able to be detected. Subsequently, comparative and evolutionary analysis can also be carried out to identify potential genetic markers, which will be useful for diagnosis and surveillance tools development in food safety.

Typically, the identified CRISPR arrays are represented by colored shapes based on nucleotide sequence identity. For facilitating and easy handling this process, Dion et al. [45] have introduced CRISPRStudio which is a user-friendly command-line tool to accelerate CRISPR analysis and standardize CRISPR array figures preparation. CRISPRStudio is able to compare nucleotide spacer sequences and then cluster them based on sequence similarity to assign a representative color; it also supports automatic sorting of CRISPR loci and highlighting shard spacers [45].

3.2 CRISPR target

In bacterial and archaeal adaptive immune systems, CRISPR-Cas targets specific protospacer nucleotide sequences in invading organisms, which requires nucleotide base pairing between processed crRNAs and target protospacer. Biswas et al. [46] have developed a flexible, interactive tool CRISPRTarget for the discovery of the target of crRNAs in diverse database. CRISPRTarget is available at http://crispr.otago.ac.nz/CRISPRTarget/crispr_analysis.html, it can be used to discover targets from both genomic and metagenomics dataset in many pathogens, including the foodborne pathogen *Salmonella*.

3.3 Conservation and diversity of *Salmonella* CRISPR arrays

Similar to other genetic components, CRISPR sequences can be conserved throughout a pathogen family. Through genomic sequence analyses of four clinically relevant *Salmonella* serovars; Enteritidis, Typhimurium, Newport and Heidelberg, it was determined that both cas operons and leaders are conserved among these four serovars [28]. Furthermore, *Salmonella* seems to be lacking in spacer acquisition, and the majority of CRISPR allelic polymorphisms usually arise from deletion or duplication of direct repeat-spacer units [47–49].

With the development of NGS technology, more and more *Salmonella* isolates have complete genome sequences available. In order to eliminate the potential bias caused by incomplete genome sequences, a collection of 115 representative *Salmonella* isolates with complete genomes (size range 4,482,117–5,395,280 bp) were analyzed in this chapter (**Table 1**). Those selected isolates come from four different subspecies, with the subspecies *enterica* as the dominant one. For the isolates within these four

ID	Subsp.	Serovar	CRISPR	Strain	Size(bp)
CP054422.1	*diarizonae*	61:k:1,5,(7)	1	14-SA00836-0	4,832,352
CP006602.1	*enterica*	4,[5],12:i:-	1,2	08-1736	4,822,189
CP034711.1	*enterica*	43:a:1,7	1,2	RSE20	4,665,063
CP007532.1	*enterica*	Abaetetuba	1,2	ATCC35640	4,547,600
CP007534.1	*enterica*	Abony	1	str.0014	4,737,447
CP049880.1	*enterica*	Adjame	1,2	381330	4,678,052
CP001138.1	*enterica*	Agona	1,2	SL483	4,798,660
CP019177.1	*enterica*	Albany	1,2	ATCC51960	4,805,448
CP007531.1	*enterica*	Anatum	1,2	ATCCBAA-1592	4,706,101
CP019116.1	*enterica*	Antsalova	1,2	S01-0511	4,648,086
CP019403.1	*enterica*	Apapa	1,2	SA20060561	4,801,658
CP019405.1	*enterica*	Bergen	1,2	ST350	4,801,835
CP030005.1	*enterica*	Berta	1,2	SA20141895	4,725,468
CP019406.1	*enterica*	Blegdam	1,2	S-1824	4,693,979
CP019407.1	*enterica*	Borreze	1,2	SA20041063	4,777,558
HF969015.2	*enterica*	Bovismorbificans	1,2	3114	4,680,283
CP022490.1	*enterica*	Braenderup	1,2	SA20026289	4,734,880
CP030002.1	*enterica*	Brandenburg	1	SA20064858	4,677,648
CP007533.1	*enterica*	Bredeney	1,2	CFSAN001080	4,603,849
CP012833.1	*enterica*	Cerro	1,2	CFSAN001588	4,651,400
AE017220.1	*enterica*	Choleraesuis	1,2	SC-B67	4,755,700
CP027677.1	*enterica*	Corvallisain	1,2	12-01738	4,887,378
CP019408.1	*enterica*	Crossness	1,2	1422-74	4,847,468
CP006055.1	*enterica*	Cubana	2	CFSAN002050	4,977,480
CP022494.1	*enterica*	Derby	1,2	SA20035215	4,850,334

ID	Subsp.	Serovar	CRISPR	Strain	Size(bp)
CP019409.1	*enterica*	Djakarta	1,2	S-1087	4,668,861
CP001144.1	*enterica*	Dublin	1,2	CT_02021853	4,842,908
NC_011294.1	*enterica*	Enteritidis	1,2	P125109	4,685,848
CP032444.1	*enterica*	Fresnoain	1,2	USMARC-69835	4,732,430
NC_011274.1	*enterica*	Gallinarum	1,2	287/91	4,658,697
*CP024165.1	*enterica*	Gaminara	1,2,3	CFSAN070644	4,801,841
CP017719.1	*enterica*	Hayindogo	1,2	CFSAN050752	4,765,719
CP001120.1	*enterica*	Heidelberg	1,2	SL476	4,888,768
CP019410.1	*enterica*	Hillingdon	1,2	N1529-D3	4,618,056
CP022503.1	*enterica*	Hvittingfoss	1	SA20014981	4,940,239
CP022015.1	*enterica*	India	1	SA20085604	5,395,280
CP022450.1	*enterica*	Indiana	1,2	D90	4,779,514
LN649235.1	*enterica*	Infantis	1,2	SINFA	4,710,675
CP019181.1	*enterica*	Inverness	1,2	ATCC10720	4,865,682
LT571437.1	*enterica*	Javaain	1	NCTC5706	4,756,780
CP004027.1	*enterica*	Javiana	1,2	CFSAN001992	4,634,161
CP019411.1	*enterica*	Johannesburg	1,2	ST203	4,651,794
CP034709.1	*enterica*	Karamoja	1,2	RSE21	4,764,896
CP022500.1	*enterica*	Kentucky	1,2	SA20030505	4,782,363
CP019412.1	*enterica*	Koessen	1	S-1501	4,566,169
CP019413.1	*enterica*	Krefeld	1,2	SA20030536	4,942,273
CP032817.1	*enterica*	Lubbock	1,2	11TTU1590	4,985,874
CP032814.1	*enterica*	Lubbock	1,2	10TTU468x	4,985,863
CP022117.1	*enterica*	Macclesfield	1,2	S-1643	4,822,139
CP019414.1	*enterica*	Manchester	1,2	ST278	4,532,753
CP022497.1	*enterica*	Manhattan	1	SA20084699	4,732,484
**CP019183.1	*enterica*	Mbandaka	1,2	ATCC51958	4,905,181
CP022489.1	*enterica*	Mbandaka	1,2	SA20026234	4,796,292
CP034713.1	*enterica*	Mikawasima	1,2	RSE15	4,650,494
CP030175.1	*enterica*	Milwaukee	1,2	SA19950795	4,822,474
CP019184.1	*enterica*	Minnesota	1,2	ATCC49284	4,592,393
CP034705.1	*enterica*	Moeroain	1,2	RSE29	4,582,521
CP007530.1	*enterica*	Montevideo	1,2	507440-20	4,694,375
CP019415.1	*enterica*	Moscow	1,2	S-1843	4,690,402
CP045056.1	*enterica*	Muenchenain	1	LG24	4,930,424
CP019201.1	*enterica*	Muenster	1,2	CFSAN001301	4,756,014
CP022663.1	*enterica*	NA	1,2	RM11065	4,991,140
CP022658.1	*enterica*	NA	1	RM11060	4,892,239

ID	Subsp.	Serovar	CRISPR	Strain	Size(bp)
CP033348.2	*enterica*	NA	1,2	CFSA1096	4,696,663
NC_011080.1	*enterica*	Newport	1,2	SL254	4,827,641
CP019416.1	*enterica*	Nitra	1,2	S-1687	4,691,807
CP022034.1	*enterica*	Onderstepoort	1,2	SA20060086	4,774,926
CP033344.1	*enterica*	Oranienburg	2	CFSAN076211	4,651,134
CP022116.1	*enterica*	Ouakam	1,2	SA20034636	4,874,915
CP012346.1	*enterica*	Panama	1	ATCC7378	4,555,576
CP000026.1	*enterica*	Paratyphi A	1,2	ATCC9150	4,585,229
CP000886.1	*enterica*	Paratyphi B	1	SPB7	4,858,887
CP000857.1	*enterica*	Paratyphi C	1,2	RKS4594	4,833,080
CP019186.1	*enterica*	Pomona	1	ATCC10729	4,482,117
CP019189.1	*enterica*	Poona	1,2	ATCCBAA-1673	4,876,720
CP012347.1	*enterica*	Pullorum	1,2	ATCC9120	4,694,842
CP022491.1	*enterica*	Saintpaul	1,2	SA20031783	4,775,303
CP001127.1	*enterica*	Schwarzengrund	2	CVM19633	4,709,075
CP029038.1	*enterica*	Senftenberg	1,2	CFSAN045763	4,766,139
CP012349.1	*enterica*	Sloterdijk	1,2	ATCC15791	4,817,791
CP017723.1	*enterica*	Stanleyville	1,2	CFSAN000624	4,888,463
CP007505.1	*enterica*	Tennessee	1,2	TXSC_TXSC08-19	4,864,410
CP006717.1	*enterica*	Thompson	1,2	RM6836	4,707,648
NC_003198.1	*enterica*	Typhi	1	CT18	4,809,037
HF937208.1	*enterica*	Typhimurium	2	DT104	4,933,631
NC_003197.2	*enterica*	Typhimurium	1,2	LT2	4,857,450
CP006048.1	*enterica*	Typhimurium var. 5	1,2	CFSAN001921	4,859,931
CP019417.1	*enterica*	Wandsworth	1,2	SA20092095	4,916,040
CP022138.1	*enterica*	Waycross	1,2	SA20041608	4,812,886
LN890520.1	*enterica*	Weltevreden,	1,2	C2346	5,129,845
CP029041.1	*enterica*	Worthington	1,2	CFSAN051295	4,914,635
CP019418.1	*enterica*	Yovokome	1	S-1850	4,640,929
CM001471.1	*houtenae*	NA	1	ATCCBAA-1581	4,672,567
CP030181.1	*NA*	NA	1,2	SA20030575	4,772,343
CP030185.1	*NA*	NA	1,2	SA20094620	4,854,398
CP030190.1	*NA*	NA	1,2	SA20104250	4,813,547
CP030196.1	*NA*	NA	1,2	SA20051401	4,869,528
CP030202.1	*NA*	NA	1,2	SA20052327	4,763,586
CP030203.1	*NA*	NA	1,2	SA20083530	5,062,813
CP030207.1	*NA*	NA	1,2	SA19992307	4,844,554
CP030209.1	*NA*	NA	1,2	SA20044414	4,805,225

ID	Subsp.	Serovar	CRISPR	Strain	Size(bp)
CP030211.1	NA	NA	1,2	SA20051528	4,719,399
CP030214.1	NA	NA	1,2	SA20025921	4,882,461
CP030217.1	NA	NA	1,2	SA20075157	4,716,530
CP030223.1	NA	NA	1	SA20083039	4,688,830
CP030225.1	NA	NA	1,2	SA20041605	4,739,617
CP030231.1	NA	NA	1,2	SA20043041	4,603,878
CP030233.1	NA	NA	2	SA20101045	4,729,786
CP030235.1	NA	NA	1,2	SA20031245	4,522,338
CP030236.1	NA	NA	1,2	SA20041606	4,524,637
CP030238.1	NA	NA	1	SA20055162	4,640,729
CP022139.1	*salamae*	55:k:z39	1,2	1315K	4,859,044
*CP029992.1	*salamae*	56:z10:e,n,x	1,2,3	SA20011914	4,807,680
*CP029995.1	*salamae*	56:b:[1, 5]	1,2,3	SA20053897	4,920,300
*CP034717.1	*salamae*	42:r:-	1,2,3	RSE09	4,860,626

**These four Salmoenlla enterica isolates contain three CRISPR arrays, which is different from the common ones containing two.*
***This isolate has the longest CRISPR2 array.*

Table 1.
Representative 115 Salmonella enterica isolates with complete genomes containing four known subspecies covering 90 serovars used for CRISPR arrays analysis in this study.

subspecies, a total of 90 different *S. enterica* serovars were included in this analysis, the details of each isolate can be found in **Table 1**.

Briefly, all available *Salmonella* complete genomes were downloaded from the NCBI database using Bioinformatics Tools (bit) (https://github.com/AstrobioMike/bit#bioinformatics-tools-bit) from GitHub and NCBI EDirect tools (https://astrobiomike.github.io/unix/ncbi_eutils). By applying common NCBI BLAST keywords, the used commands for downloading those complete genomes were:

"esearch -db assembly -query ' ("Salmonella"[Organism] OR Salmonella[All Fields]) AND (latest[filter] AND "complete genome"[filter] AND all[filter] NOT anomalous[filter])' | esummary | xtract -pattern DocumentSummary -element AssemblyAccession > Salmonella_complete_genome.txt";

"bit-dl-ncbi-assemblies -w Salmonella_complete_genome.txt -f fasta -j 12".

Then a total of 115 representative genomes were manually selected and compiled by including as many serovars as possible.

To identify CRISPR arrays in those 115 representative *Salmonella* isolates, two main used software were CRISPRDetect_2.2 (https://github.com/ambarishbiswas/CRISPRDetect_2.2) [50] and CRISPR_Studio (https://github.com/moineaulab/CRISPRStudio) [45]. Detailed procedures were described in the above two github links. Briefly, a specific python3 conda environment was created for this project, the used command for CRISPRDetect was: "*perl ../bin/CRISPRDetect_2.2/CRISPRDetect.pl -f interested.fasta -o output_file -array_quality_score_cutoff 3 -T 0*". Subsequently, the CRISPRDetect produced "output_file" containing detected CRISPR arrays was fed into and visualized using CRISPR_Studio, and the used command was: "*python*

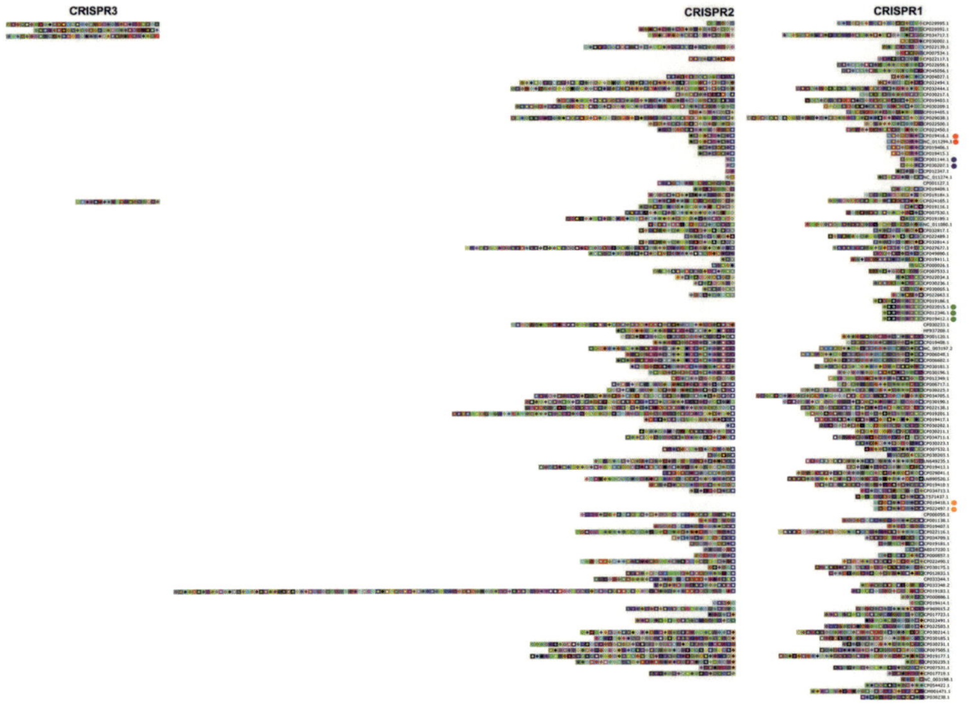

Figure 2.
Graphic representation of spacer profiles in three arrays of CRISPR1, CRISPR2 and CRIPSR3, detected from 115 Salmonella enterica isolates with complete genomes consisting of four known subspecies covering 90 serovars. The figure was created by CRISPRStudio. Each spacer is represented by a colored square and a geometric symbol. The earliest acquired spacer is shown on the right hand side and the newly acquired space is on the left hand side. Specifically, the four isolates containing CRISPR3 are: three S. enterica subsp. salamae isolates, and one S. enterica subsp. enterica serovar Gaminara, which are indicated as "" in the* **Table 1**. *The identical CRISPR spacer profiles are grouped and indicated by red, blue, green, and orange dots.*

CRISPR_Studio_1.0.py -i ../CRISPRDetect/output_file", with **Figure 2** presented in this chapter as final outputs.

Among the analyzed 115 *Salmonella* isolates, prominent diversity was observed in the detected CRIPSR array profiles. Unlike commonly reported knowledge that *Salmonella* usually contains two CRISPR loci [28], there were four isolates containing the 3rd loci, CRISPR3. Three of these isolates belonged to *S. enterica* subsp. *salamae* and the last one belonged to *S. enterica* subsp. *enterica* serovar Gaminara (**Table 1** and **Figure 2**). Additionally, there were five isolates that only contained CRISPR2 and 18 isolates that only had CRISPR1 (**Table 1** and **Figure 2**). Although prominent diversity was observed among isolates, respective identical CRISPR spacer profiles were observed for four groups. The first group (indicated by red dots) included a total of two serovars from the subspecies *enterica*, namely Nitra (CP019416) and Enteritidis (NC_011294.1); these two serovars showed high similarities and are known to be difficult to distinguish in nature using different microbiological methods. In the second group (indicated by blue dots), one was *S. enterica* subsp. *enterica* serovar Dublin (CP001144.1), and the other one is unknown serovar from the same subspecies. The third group (indicated by green dots) consisted of three serovars of *S. enterica* subsp. *enterica*, namely India (CP022015.1), Panama (CP012346.1), and Koessen (CP019412.1). The last group (indicated by orange dots) has two serovars of

Yovokome (CP019418.1) and Manhattan (CP022497.1) belonging to *S. enterica* subsp. *enterica* (**Figure 2**). The discovered CRIPSR arrays with certain similarities or dissimilarities might shed light on the phylogeny and evolutionary analysis of *Salmonella* isolates in the future.

4. Major CRISPR applications of genome editing, AMR patterns and typing tools in *Salmonella* and other microorganisms

CRISPR-Cas originates from bacteria and has also been broadly applied back in functionally studying bacteria. Here, CRISPR/Cas9 genome editing used in *Salmonella* host-pathogen interaction will be described, followed by the CRISPR-Cas diversity and its strong correlation with antimicrobial resistance (AMR) pattern studies will be introduced. The emerging application of CRISPR typing/subtyping will be explained. Finally, more advanced CRISPR-Cas related diagnosis and surveillance methods related to other microorganisms will also be demonstrated, as similar methods could be used as potential alternative methods for studying foodborne pathogens including *Salmonella*.

4.1 CRISPR/Cas9 genome editing in *Salmonella* host-pathogen interaction studies

The discovery of Cas9 has allowed for impressive advances in the field of genome editing. This protein can be utilized to modify the genome of interest, based on the segments in the CRISPR array. Cas9 endonuclease activity needs crRNAs to guarantee precise targeting, and an immediate downstream protospacer adjacent motif (PAM). With the aim of editing bacterial genomes, a vector encoding Cas9 and its guide RNAs, as well as recombination template containing required mutation are required [51]. For preventing the re-cleavage of Cas9 of the target genome, the spacer of PAM sequences will need to be modified [52, 53]. Using such approach, mutations have been introduced into the *sdiA* gene to study its effect on *S. enterica* pathogenesis. The introduced mutations affected *S. enterica* biofilm formation, cell adhesion and invasion [54]. CRISPR/Cas9 has also been used in generating macrophage knockout mice cell lines, which facilitates *S.* Typhimurium infection studies by determining the contribution of background contaminations in the phenotypes of primary macrophages from congenic mice [55]. It has also demonstrated that the CRISPR-Cas system is involved in the resistance to bile salts and biofilm formation in *S.* Typhi [29]. This demonstrated CRISPR/Cas9 genome editing based methods contribute significantly in carrying out functional studies of *Salmonella.*

4.2 CRISPR/Cas diversity and its strong correlation with AMR pattern

AMR is a global concern for human health and a World health organization global priority (https://www.who.int/news-room/fact-sheets/detail/antimicrobial-resistance). WGS is replacing traditional phenotypic method such as disk diffusion method for routine testing of foodborne pathogens AMR. The tools of ResFinder [56, 57] or the Comprehensive Antibiotic Resistance Database (CARD) [58] detect the presence of AMR genes in an isolate by comparing its sequence against known genes cataloged in a reference database of known AMR determinants. Although knowledge of the CRISPR-Cas systems has been applied in many research areas, there are not many studies in applying it to the analysis of antibiotic resistance in *Salmonella*. Recently, by using large-scale bioinformatics investigation of the 1059 isolates of *S.* Typhi CRISPR-Cas systems, 47 unique spacers and 15 unique DRs were identified, as well as unique conservation and clonality of the *S.* Typhi type I-E CRISPR-Cas system was observed [59]. The identified spacers and repeats showed specific patterns which demonstrated significant associations with AMR status, genotype, and demographic characteristics. This suggests they have the potential to be used as biomarkers to develop rapid and inexpensive diagnostics tests [59]. Similarly, on Chinese poultry farms, analysis of 75 *Salmonella isolates* consisting of 11 serovars, found that there were close correlations between CRISPR loci and AMRs, however, there was no close correlations between CRISPR loci and antibiotics [60].

4.3 CRISPR typing and subtyping as improved laboratory diagnostic tool in *Salmonella*

Various molecular and phenotypic typing techniques have been developed to track bacterial origins, for instance, pulse-field gel electrophoresis (PFGE), phage typing, multi-locus sequence typing (MLST), multi-locus variable number tandem repeat (MLVA) and single nucleotide polymorphism (SNP) pipelines [61]. The above mentioned typing methods are limited in both speed and precision. In recent years, improved and innovative surveillance tools of CRISPR typing have been developed, which are used to gain knowledge in better understanding a variety of bacteria, such as *Salmonella* [47]. Serving as a complementary tool for the high-resolution core genome single nucleotide variant (cgSNV) method, CRISPR typing was useful for determining source attribution in foodborne *S.* Heidelberg outbreaks [62, 63]. CRISPR typing was also shown to facilitate further studies in understanding the virulence and global distribution of the *S.* Virchow serovar [64]. Furthermore, the combination of both MLVA and CRISPR (CRISPR-MLVA method) gave better genotyping results than using each one alone, when testing 171 *Salmonella* strains from nine serovars [48]. There are limitations to this method of typing, particularly in very closely related isolates. In these instances, it has been shown that using CRISPR typing in conjunction with a SNP analysis allows for better resolution, indicating the use of CRISPR typing still exhibits clear benefits [65]. A few CRISPR based typing tools are illustrated in details as below.

4.3.1 Conventional CRISPR typing

In conventional CRISPR typing (CCT), all spacer sequences in the two loci of CRISPR1 and CRISPR2 are extracted [48]. Then CRISPR1 and CRISPR2 spacer sequences profiles are analyzed and visualized using CRIPSRviz [66]. There are three main procedures: (1) CRISPR arrays are obtained by either directed whole-genome sequencing or PCR amplification of CRISPR loci using conservative sequences following by sequencing; (2) the identification and characterization of potential CRISPR arrays based on the previous sequencing results; (3) finally, clustering of analyzed isolates based on the absence or presence of analyzed CRISPR arrays.

4.3.2 CRISPR locus spacer pair typing

CCT can be labor intensive to carry out. To increase the easy of typing, CRISPR locus spacer pair typing (CLSPT) was developed [67]. Instead of using all the obtained spacer sequences, only one spacer sequence in both CRISPR1 and CRISPR2 loci will be used for typing. This spacer sequence is the first one, found closest to the leader sequences. In this method, the first spacer sequence of the CRISPR1 leader sequences is combined with the first spacer sequence of the CRISPR2 leader sequences. Then these two spacer sequences were used as the total sequences for *Salmonella* strain typing.

4.3.3 CRISPR locus three spacer sequences typing

Usually, during the evolution of bacterial strains which contain CRISPR arrays, the first captured exogenous nucleotide sequence could display strain origins, and the spacers from the same serotype possess certain conservation. Thus, Li et al. have developed a *Salmonella* typing method, called CRISPR locus three spacer sequences typing (CLTSST) method, which could be used to distinguish different serotype clusters. They used three spacer sequences including the initial two spacer sequences (the first acquisitions or the ones with the furthest distance to the leader sequence) and latest spacer sequence close to the leader sequence are combined and used as the total analyzed sequences for strain typing [60].

4.3.4 Conserved CRISPR arrays serving as quantitative PCR targets

In addition to the sequences analyses of CRISPR loci typing in *Salmonella*, the conserved CRISPR arrays can also be used as targets for qPCR primers and probes design. It has been demonstrated that a *S.* Infantis-specific qPCR assay is able to detect the Infantis serovar from mixed cultures of *Salmonella* down to 0.1% of the population, and with the detection sensitivity of 10 colony forming units [21]. For the utility of this CRIPSR based qPCR molecular approach in improved surveillance system, two main parameters need to be met in regards to the CRISPR spacer sequences that are to be used for designing primers and probes: (1) the used spacers need to be specific for the tested serovar; (2) the selected spacers need to be conserved and present in all strains of that specific serovar.

4.3.5 Other related applications of CRISPR-Cas

It is well-known that efficient delivery of a CRISPR/Cas9 plasmid is critical for effective therapy in clinical settings. Other than a receipt of a plasmid carrying CRISPR/Cas, it has also been found that *Salmonella* can be used as a CRISPR/Cas9 plasmid carrier for *in vivo* therapy against virus-induced cancer [68]. It has been demonstrated that the usage of *Salmonella* in CRISPR system provides a simpler and more effective platform for *in vivo* therapy [68].

The CRISPR-Cas system can also be used for diagnostics by utilizing the properties of the proteins themselves. For example, the Cas9 protein has been used in detection assays to help increase the percentage of the genomic regions of interest that is present. During library preparation in NGS projects, a method known as Depletion of Abundant Sequences by Hybridization uses recombinant Cas9 protein complexed with a library of guide RNAs to target and cleave unwanted DNA, leading to the increased yield of sequences of interest [69].

Additionally, the method Finding Low Abundance Sequences by Hybridization (FLASH) uses Cas9 and guide RNAs that allow sequences of interest to be cleaved into an ideal size for NGS sequencing, increasing the presence of reads that can be captured from the sequence of interest [70]. This can be used for example to detect antimicrobial resistant populations that may be present in low levels compared to the wild type. Recently, a modified Cas9 variant has been developed (SpCas9 named SpRY), that allows for the digestion of specific regions without the requirement of the PAM sequence [71].

Other diagnostic tools have also been developed with the CRISPR-Cas system. Although the most reported cas protein in *Salmonella* is related to Cas9, the other two most popular cas proteins related to diagnosis tests are *cas12a* and *cas13*. Different from Cas9 which cuts double stranded DNA relying on a precise location "T rich" PAM, both *cas12a* [72] and *cas13* [73] remain bound to the target and then cleave other DNA/RNA non-discriminately. This feature is recognized as collateral cleavage of trans-cleavage activity, which has been broadly applied in the development of various diagnostic technologies [74]. DNA endonuclease-targeted CRISPR trans reporter is a method developed using the Cas12a protein and its ability to degrade single-stranded DNA (ssDNA). Using this property, along with a ssDNA reporter, this system can be used to detect if specific pathogen types are present in a sample [72]. Another example can be seen from the global pandemic associated with betacoronavirus severe acute respiratory syndrome coronavirus 2 (SARS-CoV-2). For the detection of SARS-CoV-2 in clinical validations, specific high-sensitivity enzymatic reporter unlocking (SHERLOCK), has been shown to be a promising alternative method to qPCR with regards to its visualization speed and experimental settings with limited resources [75–77]. These are just a few ways the CRISPR-Cas system can be alternatively used for diagnostic purposes. These diagnostic or detection methods could also be adapted and used as alternative methods for surveillance or typing in *Salmonella*.

5. Conclusions

The CRISPR-Cas system in *Salmonella* has been shown to be useful in differentiating between different strains. According to the WGS based genome analysis of 115 isolates, three *S. enterica* subsp. *salamae* isolates, and one *S. enterica* subsp. *enterica* serovar Gaminara possess three CRISPR loci, namely CRISPR1, CRISPR2 and CRISPR3, which differs from the commonly reported two CRIPR loci CRISPR1 and CRISPR2 in *Salmonella*. On the contrary, 18 isolates only had CRISPR1 and five isolates only had CRISPR2. With the emerging applications of CRISPR arrays in *Salmonella* genome editing, AMR studies, typing and subtyping in diagnosis and surveillance, a thorough investigation of the uses of CRISPR-Cas will facilitate better understanding its host-pathogen interaction, immune response and its usages in improving laboratory tests. Adapting the many advanced CRIPSR-based diagnostic tools such as SHERLOCK, and FLASH, will allow for faster detection and/or the ability for more detailed analyses to be carried out. This will allow for improved laboratory diagnosis and surveillance endeavors in food safety, as well as offer better tools for any future outbreak responses.

Acknowledgements

This work was supported by the Public Health Agency of Canada and Canadian Food Inspection Agency.

Conflict of interest

The authors declares no conflict of interest.

Acronyms and abbreviations

Acr	anti-CRISPR protein
AMR	antimicrobial resistance
CARD	comprehensive Antibiotic Resistance Database
CRISPR	clustered regularly interspaced short palindromic repeats
CCT	conventional CRISPR typing
CLSPT	CRISPR locus spacer Pair Typing
CLTSST	CRISPR Locus Three Spacer Sequences Typing
crRNAs	CRISPR RNAs
DR	direct repeats
FLASH	Finding Low Abundance Sequences by Hybridization
MLVA	multi-locus variable number tandem repeat
MLST	multi-locus sequence typing
NCBI	National Center for Biotechnology Information
NTS	non-typhoidal *Salmonella*
NGS	next generation sequencing
PAM	protospacer adjacent motif
PFGE	pulsed-field gel electrophoresis
SARS-CoV-2	severe acute respiratory syndrome coronavirus 2
SHERLOCK	specific high-sensitivity enzymatic reporter unlocking
ssDNA	single-stranded DNA
SNP	single nucleotide polymorphism
WGS	whole genome sequencing

Author details

Ruimin Gao[1,2]* and Jasmine Rae Frost[1]

1 National Microbiology Laboratory, Public Health Agency of Canada, Winnipeg, Manitoba, Canada

2 Ottawa Laboratory Fallowfield, Canadian Food Inspection Agency, Ottawa, Ontario, Canada

*Address all correspondence to: ruimin.gao@phac-aspc.gc.ca

References

[1] Barrangou R, Fremaux C, Deveau H, Richards M, Boyaval P, Moineau S, et al. CRISPR provides acquired resistance against viruses in prokaryotes. Science. 2007;**315**(5819):1709-1712

[2] Medina-Aparicio L, Davila S, Rebollar-Flores JE, Calva E, Hernandez-Lucas I. The CRISPR-Cas system in *Enterobacteriaceae*. Pathogens and Disease. 2018;**76**(1)

[3] Makarova KS, Wolf YI, Alkhnbashi OS, Costa F, Shah SA, Saunders SJ, et al. An updated evolutionary classification of CRISPR-Cas systems. Nature Reviews Microbiology. 2015;**13**(11):722-736

[4] Wright AV, Nunez JK, Doudna JA. Biology and applications of CRISPR systems: Harnessing nature's toolbox for genome engineering. Cell. 2016;**164**(1-2):29-44

[5] Westra ER, van Houte S, Gandon S, Whitaker R. The ecology and evolution of microbial CRISPR-Cas adaptive immune systems. Philosophical Transactions of the Royal Society of London. Series B, Biological Sciences. 2019;**374**(1772):20190101

[6] Chylinski K, Makarova KS, Charpentier E, Koonin EV. Classification and evolution of type II CRISPR-Cas systems. Nucleic Acids Research. 2014;**42**(10):6091-6105

[7] Makarova KS, Wolf YI, Iranzo J, Shmakov SA, Alkhnbashi OS, Brouns SJJ, et al. Evolutionary classification of CRISPR-Cas systems: a burst of class 2 and derived variants. Nature Reviews Microbiology. 2020;**18**(2):67-83

[8] Makarova KS, Zhang F, Koonin EV. SnapShot: Class 1 CRISPR-Cas Systems. Cell. 2017;**168**(5):946-946 e941

[9] Makarova KS, Zhang F, Koonin EV. SnapShot: Class 2 CRISPR-Cas Systems. Cell. 2017;**168**(1-2):328-328 e321

[10] Xu C, Zhou Y, Xiao Q, He B, Geng G, Wang Z, et al. Programmable RNA editing with compact CRISPR-Cas13 systems from uncultivated microbes. Nature Methods. 2021;**18**(5):499-506

[11] Touchon M, Rocha EP. The small, slow and specialized CRISPR and anti-CRISPR of *Escherichia* and *Salmonella*. PLoS One. 2010;**5**(6):e11126

[12] Makarova KS, Haft DH, Barrangou R, Brouns SJ, Charpentier E, Horvath P, et al. Evolution and classification of the CRISPR-Cas systems. Nature Reviews Microbiology. 2011;**9**(6):467-477

[13] Hille F, Charpentier E. CRISPR-Cas: biology, mechanisms and relevance. Philosophical Transactions of the Royal Society of London. Series B, Biological Sciences. 2016;**371**(1707)

[14] Garneau JE, Dupuis ME, Villion M, Romero DA, Barrangou R, Boyaval P, et al. The CRISPR/Cas bacterial immune system cleaves bacteriophage and plasmid DNA. Nature. 2010;**468**(7320):67-71

[15] Yosef I, Goren MG, Qimron U. Proteins and DNA elements essential for the CRISPR adaptation process in *Escherichia coli*. Nucleic Acids Research. 2012;**40**(12):5569-5576

[16] Nunez JK, Kranzusch PJ, Noeske J, Wright AV, Davies CW, Doudna JA. Cas1-Cas2 complex formation mediates spacer acquisition during CRISPR-Cas adaptive immunity. Nature Structural & Molecular Biology. 2014;**21**(6):528-534

[17] Deltcheva E, Chylinski K, Sharma CM, Gonzales K, Chao Y, Pirzada ZA, et al. CRISPR RNA maturation by trans-encoded small RNA and host factor RNase III. Nature. 2011;**471**(7340):602-607

[18] Shariat N, Dudley EG. CRISPRs: molecular signatures used for pathogen subtyping. Applied and Environmental Microbiology. 2014;**80**(2):430-439

[19] Garcia-Gutierrez E, Almendros C, Mojica FJ, Guzman NM, Garcia-Martinez J. CRISPR content correlates with the pathogenic potential of *Escherichia coli*. PLoS One. 2015;**10**(7):e0131935

[20] Delannoy S, Beutin L, Fach P. Use of clustered regularly interspaced short palindromic repeat sequence polymorphisms for specific detection of enterohemorrhagic *Escherichia coli* strains of serotypes O26:H11, O45:H2, O103:H2, O111:H8, O121:H19, O145:H28, and O157:H7 by real-time PCR. Journal of Clinical Microbiology. 2012;**50**(12):4035-4040

[21] Richards AK, Hopkins BA, Shariat NW. Conserved CRISPR arrays in *Salmonella enterica* serovar Infantis can serve as qPCR targets to detect Infantis in mixed serovar populations. Letters in Applied Microbiology. 2020;**71**(2):138-145

[22] The Royal Swedish Academy of Sciences has Awarded the Nobel Prize in Chemistry 2020 to Emmanuelle Charpentierthe and Jennifer A. Doudna. Press Release. 2020

[23] Ledford H. CRISPR, the disruptor. Nature. 2015;**522**(7554):20-24

[24] NPR.org. In a 1st, Doctors in U.S. use CRISPR tool to treat patient with genetic disorder. 2019. Available from: https://www.npr.org/sections/health-shots/2019/07/29/744826505/sickle-cell-patient-reveals-why-she-is-volunteering-for-landmark-gene-editing-st

[25] National Public Radio Organization. In a 1st, Scientists Use Revolutionary Gene-Editing Tool to Edit Inside a Patient. Health News from NPR. 2020

[26] Heler R, Marraffini LA, Bikard D. Adapting to new threats: the generation of memory by CRISPR-Cas immune systems. Molecular Microbiology. 2014;**93**(1):1-9

[27] Horvath P, Barrangou R. CRISPR/Cas, the immune system of bacteria and archaea. Science. 2010;**327**(5962):167-170

[28] Shariat N, Timme RE, Pettengill JB, Barrangou R, Dudley EG. Characterization and evolution of *Salmonella* CRISPR-Cas systems. Microbiology (Reading). 2015;**161** (Pt 2):374-386

[29] Medina-Aparicio L, Rodriguez-Gutierrez S, Rebollar-Flores JE, Martinez-Batallar AG, Mendoza-Mejia BD, Aguirre-Partida ED, et al. The CRISPR-cas system is involved in ompr genetic regulation for outer membrane protein synthesis in *Salmonella* Typhi. Frontiers in Microbiology. 2021;**12**:657404

[30] Pettengill JB, Timme RE, Barrangou R, Toro M, Allard MW, Strain E, et al. The evolutionary history and diagnostic utility of the CRISPR-Cas system within *Salmonella enterica* ssp. *enterica*. PeerJ. 2014;**2**:e340

[31] Rauch BJ, Silvis MR, Hultquist JF, Waters CS, McGregor MJ, Krogan NJ, et al. Inhibition of CRISPR-Cas9 with bacteriophage proteins. Cell. 2017;**168** (1-2):150-158 e110

[32] Bondy-Denomy J, Pawluk A, Maxwell KL, Davidson AR. Bacteriophage genes that inactivate the CRISPR/Cas bacterial immune system. Nature. 2013;**493**(7432):429-432

[33] Maxwell KL, Garcia B, Bondy-Denomy J, Bona D, Hidalgo-Reyes Y, Davidson AR. The solution structure of an anti-CRISPR protein. Nature Communications. 2016;7:13134

[34] Di H, Ye L, Yan H, Meng H, Yamasak S, Shi L. Comparative analysis of CRISPR loci in different *Listeria monocytogenes* lineages. Biochemical and Biophysical Research Communications. 2014;**454**(3):399-403

[35] Sesto N, Touchon M, Andrade JM, Kondo J, Rocha EP, Arraiano CM, et al. A PNPase dependent CRISPR system in *Listeria*. PLoS Genetics. 2014;**10**(1):e1004065

[36] Fuchsbauer O, Swuec P, Zimberger C, Amigues B, Levesque S, Agudelo D, et al. Cas9 allosteric inhibition by the anti-CRISPR protein AcrIIA6. Molecular Cell. 2019;**76**(6):922-937 e927

[37] Mottawea W, Duceppe MO, Dupras AA, Usongo V, Jeukens J, Freschi L, et al. *Salmonella enterica* prophage sequence profiles reflect genome diversity and can be used for high discrimination subtyping. Frontiers in Microbiology. 2018;**9**:836

[38] Banerji S, Simon S, Tille A, Fruth A, Flieger A. Genome-based *Salmonella* serotyping as the new gold standard. Scientific Reports. 2020;**10**(1):4333

[39] Sorek R, Kunin V, Hugenholtz P: CRISPR -- a widespread system that provides acquired resistance against phages in bacteria and archaea. Nature Reviews Microbiology. 2008;**6**(3):181-186

[40] Grissa I, Vergnaud G, Pourcel C. CRISPRFinder. a web tool to identify clustered regularly interspaced short palindromic repeats. Nucleic Acids Research. 2007;**35**(Web Server issue):W52-W57

[41] Grissa I, Vergnaud G, Pourcel C. The CRISPRdb database and tools to display CRISPRs and to generate dictionaries of spacers and repeats. BMC Bioinformatics. 2007;**8**:172

[42] Bland C, Ramsey TL, Sabree F, Lowe M, Brown K, Kyrpides NC, et al. CRISPR recognition tool (CRT): a tool for automatic detection of clustered regularly interspaced palindromic repeats. BMC Bioinformatics. 2007;**8**:209

[43] Grissa I, Vergnaud G, Pourcel C. CRISPRcompar: a website to compare clustered regularly interspaced short palindromic repeats. Nucleic Acids Research. 2008;**36**(Web Server issue):W145-W148

[44] Couvin D, Bernheim A, Toffano-Nioche C, Touchon M, Michalik J, Neron B, et al. CRISPRCasFinder, an update of CRISRFinder, includes a portable version, enhanced performance and integrates search for Cas proteins. Nucleic Acids Research. 2018;**46**(W1):W246-W251

[45] Dion MB, Labrie SJ, Shah SA, Moineau S. CRISPRStudio: A user-friendly software for rapid CRISPR array visualization. Viruses. 2018;**10**(11)

[46] Biswas A, Gagnon JN, Brouns SJ, Fineran PC, Brown CM. CRISPRTarget: bioinformatic prediction and analysis of crRNA targets. RNA Biology. 2013;**10**(5):817-827

[47] Fabre L, Zhang J, Guigon G, Le Hello S, Guibert V, Accou-Demartin M,

et al. CRISPR typing and subtyping for improved laboratory surveillance of *Salmonella* infections. PLoS One. 2012;7(5):e36995

[48] Liu F, Barrangou R, Gerner-Smidt P, Ribot EM, Knabel SJ, Dudley EG. Novel virulence gene and clustered regularly interspaced short palindromic repeat (CRISPR) multilocus sequence typing scheme for subtyping of the major serovars of *Salmonella enterica* subsp. *enterica*. Applied and Environmental Microbiology. 2011;77(6):1946-1956

[49] Shariat N, DiMarzio MJ, Yin S, Dettinger L, Sandt CH, Lute JR, et al. The combination of CRISPR-MVLST and PFGE provides increased discriminatory power for differentiating human clinical isolates of *Salmonella enterica* subsp. *enterica* serovar Enteritidis. Food Microbiology. 2013;**34**(1):164-173

[50] Biswas A, Staals RH, Morales SE, Fineran PC, Brown CM. CRISPRDetect: a flexible algorithm to define CRISPR arrays. BMC Genomics. 2016;**17**:356

[51] Jiang W, Bikard D, Cox D, Zhang F, Marraffini LA. RNA-guided editing of bacterial genomes using CRISPR-Cas systems. Nature Biotechnology. 2013;**31**(3):233-239

[52] Jiang Y, Chen B, Duan C, Sun B, Yang J, Yang S. Multigene editing in the *Escherichia coli* genome via the CRISPR-Cas9 system. Applied and Environmental Microbiology. 2015;**81**(7):2506-2514

[53] Huang H, Zheng G, Jiang W, Hu H, Lu Y. One-step high-efficiency CRISPR/Cas9-mediated genome editing in *Streptomyces*. Acta Biochimica et Biophysica Sinica Shanghai. 2015;**47**(4):231-243

[54] Askoura M, Almalki AJ, Lila ASA, Almansour K, Alshammari F, Khafagy ES, et al. Alteration of *Salmonella enterica* virulence and host pathogenesis through targeting *sdiA* by using the CRISPR-Cas9 system. Microorganisms. 2021;**9**(12)

[55] Ferrand J, Croft NP, Pepin G, Diener KR, Wu D, Mangan NE, et al. The use of CRISPR/Cas9 gene editing to confirm congenic contaminations in host-pathogen interaction studies. Frontiers in Cellular and Infection Microbiology. 2018;**8**:87

[56] Camacho C, Coulouris G, Avagyan V, Ma N, Papadopoulos J, Bealer K, et al. BLAST+: architecture and applications. BMC Bioinformatics. 2009;**10**:421

[57] Clausen P, Aarestrup FM, Lund O. Rapid and precise alignment of raw reads against redundant databases with KMA. BMC Bioinformatics. 2018;**19**(1):307

[58] Alcock BP, Raphenya AR, Lau TTY, Tsang KK, Bouchard M, Edalatmand A, et al. CARD 2020: antibiotic resistome surveillance with the comprehensive antibiotic resistance database. Nucleic Acids Research. 2020;**48**(D1):D517-D525

[59] Tanmoy AM, Saha C, Sajib MSI, Saha S, Komurian-Pradel F, van Belkum A, et al. CRISPR-Cas diversity in clinical *Salmonella enterica* serovar Typhi isolates from south Asian countries. Genes (Basel). 2020;**11**(11)

[60] Li C, Wang Y, Gao Y, Li C, Ma B, Wang H. Antimicrobial resistance and CRISPR typing among *Salmonella* isolates from poultry farms in China. Frontiers in Microbiology. 2021;**12**:730046

[61] Ogunremi D, Gao R, Slowey R, Chen S, Andrievskaia O, Bekal S, et al. Tracking *Salmonella* Enteritidis in the genomics era: clade definition using a SNP-PCR assay and implications for population structure. In: Alexandre, editor. *Salmonella App A Global*

Challenge. Lamas PRaCMF: London, UKIntechOpen; 2021

[62] Yousfi K, Usongo V, Berry C, Khan RH, Tremblay DM, Moineau S, et al. Source tracking based on core genome SNV and CRISPR typing of *Salmonella enterica* serovar Heidelberg isolates involved in foodborne outbreaks in Quebec, 2012. Frontiers in Microbiology. 2020;**11**:1317

[63] Vincent C, Usongo V, Berry C, Tremblay DM, Moineau S, Yousfi K, et al. Comparison of advanced whole genome sequence-based methods to distinguish strains of *Salmonella enterica* serovar Heidelberg involved in foodborne outbreaks in Quebec. Food Microbiology. 2018;**73**:99-110

[64] Bachmann NL, Petty NK, Ben Zakour NL, Szubert JM, Savill J, Beatson SA. Genome analysis and CRISPR typing of *Salmonella enterica* serovar Virchow. BMC Genomics. 2014;**15**:389

[65] Nadin-Davis S, Pope L, Devenish J, Allain R, Ogunremi D. Evaluation of the use of CRISPR loci for discrimination of *Salmonella enterica* subsp. *enterica* serovar Enteritidis strains recovered in Canada and comparison with other subtyping methods. AIMS Microbiology. 2022;**8**(3):300-317

[66] Nethery MA, Barrangou R. CRISPR Visualizer: rapid identification and visualization of CRISPR loci via an automated high-throughput processing pipeline. RNA Biology. 2019;**16**(4):577-584

[67] Li H, Li P, Xie J, Yi S, Yang C, Wang J, et al. New clustered regularly interspaced short palindromic repeat locus spacer pair typing method based on the newly incorporated spacer for *Salmonella enterica*. Journal of Clinical Microbiology. 2014;**52**(8):2955-2962

[68] Senevirathne A, Hewawaduge C, Lee JH. Genetic interference exerted by *Salmonella*-delivered CRISPR/Cas9 significantly reduces the pathological burden caused by Marek's disease virus in chickens. Veterinary Research. 2021;**52**(1):125

[69] Gu W, Crawford ED, O'Donovan BD, Wilson MR, Chow ED, Retallack H, et al. Depletion of abundant sequences by hybridization (DASH): using Cas9 to remove unwanted high-abundance species in sequencing libraries and molecular counting applications. Genome Biology. 2016;**17**:41

[70] Quan J, Langelier C, Kuchta A, Batson J, Teyssier N, Lyden A, et al. FLASH: a next-generation CRISPR diagnostic for multiplexed detection of antimicrobial resistance sequences. Nucleic Acids Research. 2019;**47**(14):e83

[71] Walton RT, Christie KA, Whittaker MN, Kleinstiver BP. Unconstrained genome targeting with near-PAMless engineered CRISPR-Cas9 variants. Science. 2020;**368**(6488):290-296

[72] Chen JS, Ma E, Harrington LB, Da Costa M, Tian X, Palefsky JM, et al. CRISPR-Cas12a target binding unleashes indiscriminate single-stranded DNase activity. Science. 2018;**360**(6387):436-439

[73] Abudayyeh OO, Gootenberg JS, Konermann S, Joung J, Slaymaker IM, Cox DB, et al. C2c2 is a single-component programmable RNA-guided RNA-targeting CRISPR effector. Science. 2016;**353**(6299):aaf5573

[74] Gootenberg JS, Abudayyeh OO, Kellner MJ, Joung J, Collins JJ, Zhang F. Multiplexed and portable nucleic acid detection platform with

Cas13, Cas12a, and Csm6. Science. 2018;**360**(6387):439-444

[75] Broughton JP, Deng X, Yu G, Fasching CL, Servellita V, Singh J, et al. CRISPR-Cas12-based detection of SARS-CoV-2. Nature Biotechnology. 2020;**38**(7):870-874

[76] Patchsung M, Jantarug K, Pattama A, Aphicho K, Suraritdechachai S, Meesawat P, et al. Clinical validation of a Cas13-based assay for the detection of SARS-CoV-2 RNA. Nature Biomedical Engineering. 2020;**4**(12):1140-1149

[77] Joung J, Ladha A, Saito M, Kim NG, Woolley AE, Segel M, et al. Detection of SARS-CoV-2 with SHERLOCK one-pot testing. The New England Journal of Medicine. 2020;**383**(15):1492-1494

Chapter 5

Perspective Chapter: Solar Disinfection – Managing Waterborne *Salmonella* Outbreaks in Resource-Poor Communities

Cornelius Cano Ssemakalu

Abstract

Salmonella outbreaks remain a significant problem in many resource-poor communities globally, especially in low and middle-income countries (LMICs). These communities cannot reliably access treated piped water, thus reverting to the use of environmental water for domestic and agricultural purposes. In most LMICs, the maintenance and expansion of the existing wastewater and water treatment infrastructure to meet the growing population are not considered. This results in regular wastewater and water treatment failures causing an increase in an assortment of waterborne pathogens, including *Salmonella*. Solving these problems would require the maintenance, expansion and construction of new wastewater and water treatment infrastructure. The implementation of such interventions would only occur over a long period. Unfortunately, time is not a luxury in communities experiencing the effects of such problems. However, highly disruptive household interventions such as solar disinfection (SODIS) could be implemented in communities experiencing endemic *Salmonella* outbreaks. SODIS has been shown to inactivate a variety of water-related pathogens. SODIS requires significantly less financial input to implement in comparison to other household-level interventions. Various studies have shown better health outcomes due to SODIS in communities that previously struggled with waterborne diseases, including *Salmonella*. The aim of this chapter is to share a perspective on the continued reliance on SODIS as for the control waterborne *Salmonella* in LMICs.

Keywords: SODIS, *Salmonella*, sanitation, hygiene, water treatment, disinfection, filtration, Coagulation, Flocculation, oxidation, water, Waterborne, LMIC

1. Introduction

The genus *Salmonella* consists of two species with over 2500 serovars. The serovars within the species *Salmonella enterica* are classified as either typhoidal or non-typhoidal. Although genetically similar these serovars elicit significantly different diseases. Typhoidal *Salmonella* serovars such as Typhi and Paratyphi A are human restricted and cause an invasive systemic typhoid fever that is life threating in both healthy and

IntechOpen

immune compromised individuals [1]. Non typhoidal *Salmonella* (NTS) serovars such as Typhimurium and Enteritidis cause self-limiting gastroenteritis in either humans or animals. The gastroenteritis caused by NTS is often mild in healthy adults but severe in immune compromised individuals [2, 3].

S. enterica infections primarily those associated with serovars Typhimurium and Enteritidis [4, 5] remain a global burden especially in low and middle income countries in Africa and Asia affecting more than 93 million people and causing the deaths of over 1.2 million people globally [6, 7]. Most of these infections and deaths occurred in people living in resource-poor communities, especially those in Low- and Middle-Income Countries (LMICs) in Africa [4, 8]. This could be attributed to the high prevalence of malnutrition and immune compromising diseases such as malaria and AIDS. Infections due to *Salmonella* are not exclusive to LMICs. According to the Centre for Disease Control (CDC), more than 1 million people in the United States of America (USA) experience a *Salmonella* infection. This costs the USA more than $ 3.7 billion US in medical costs [9]. Recently, the World Health Organisation (WHO) was alerted to a *Salmonella* outbreak associated with European food products for the European and Global markets [10]. The CDC estimates that 46% of foodborne diseases and 23% of deaths are linked to produce consumption [9]. In High-Income Countries (HIC), there is a more likelihood of acquiring a *Salmonella* infection through the consumption of food products as opposed to water [11, 12].

This chapter highlights the role played by water in the transmission of *Salmonella* especially in resource poor LMICs. Thereafter, an overview of how water and sanitation infrastructure is prioritised in Africa is provided. This is followed by an evaluation of water treatment approaches that could be used at a household level to reduce the burden of *Salmonella*. The chapter ends by providing reasons why SODIS is an ideal water treatment intervention at a household level.

2. Water and *Salmonella* infections

Environmental water resources play a critical role in food crop produce linked to *Salmonella* infections occurring in HICs [12, 13]. Environmental water bodies can harbour *Salmonella* for several months [13, 14]. This makes *Salmonella* a waterborne pathogen that could be introduced into a susceptible animal host when untreated environmental water is consumed or used for domestic and agricultural purposes. Previously, *Salmonella* infections were mainly associated with consuming contaminated animal products. However, in recent years, *Salmonella* outbreaks associated with consuming contaminated food crops such as fresh fruits, vegetables, spices, and nuts have increased [15, 16]. This is probably driven by the increased adoption of a vegetarian or vegan lifestyle [17]. The presence of *Salmonella* on food crops has been attributed to the microbiological quality of water used for irrigation [12, 16].

Water remains a key factor in the transmission of *S. enterica* in LMICs [14] and HICs [12, 13, 18]. Access to clean water is a fundamental human right. However, many resource-poor communities worldwide, especially LMICs, struggle to access clean water [14]. Currently, more than 2 billion people lack access to safely managed water, of which more than 700 million live without basic drinking water. Most of these live in Africa [19]. The current paradigm of *Salmonella* infection places poor sanitary habits and practices as critical contributors toward the reintroduction of *S. enterica* into the environmental water resources. Although an increase in global sanitation has been reported, more than 3.6 billion people lack access to well-managed sanitation, of which 1.7 billion still lack basic sanitation [19]. Therefore, people living in resource-poor communities in LMICs contract *Salmonella* infections by consuming contaminated food and water [14]. But, if these communities had access to treated water, a reduction in waterborne *Salmonella* would occur as observed in HICs.

Furthermore, practicing proper sanitation and hygiene in tandem with the availability of treated water would reduce the prevalence of *S. enterica* in the environmental water resources. This would improve the microbiological quality of natural water resources for agricultural purposes. Providing resource-poor communities with clean water would require establishing effective sanitary and water treatment infrastructure.

3. Investment in water and sanitation infrastructure with focus on Africa

Sanitary and water treatment infrastructure availability is a major driver of economic development because it curbs health risks, enables education and other productive activities, and enhances the labour force's productivity [20, 21]. For instance, the lack of proper sanitation in South Asia results in financial and economic losses of up to 2 and 9 billion dollars, respectively, while adequate sanitation infrastructure in France enables tourism and sustains the jobs of more than 2000 people in the tourism sector [20].

The African continent consists of 53 member states with a combined population of more than 1.4 billion people [22]. Currently, the African continent has the highest population below the age of 15 [23]. By 2050 Africa will be home to more than half of the world's population, and 1.3 billion people will live in urbanised areas [24, 25]. About 56% of people living in urban areas in Africa have access to piped water compared to 67% in 2003, and just 11% can access a sewer connection [26]. This observation implies that the current infrastructure cannot support the increasing population and hence threatens social stability and may act as a driver of migration within and out of Africa [26]. Given the growing population, it is logical to prioritise the expansion of existing as well as the construction of new sanitary and water treatment infrastructure in Africa.

However, this is not reflected in the African infrastructural commitments. In 2017 the transport infrastructure sector received the highest commitment ($ 34 billion, 41.7% of the total obligations) in comparison to the water infrastructure sector ($ 13.2 billion, 16.2% of the total commitments) [27]. In the same year, the funding gap for transport infrastructure (8%) was lower than that of the water infrastructure (84%) sector [27]. Previous reports showed that the transport and water sectorial infrastructural commitments had increased by 30 and 8% between 2016 and 2017 [27]. Nevertheless, African states' water and sanitation infrastructure financing declined by 3% between 2016 and 2017 [27]. During that period, foreign aid commitments were made to finance water and sanitation infrastructural projects in many Low-Income Countries (LIC) in Africa. For instance, Italy committed $ 69 million to a Mozambique water and sanitation project. China committed $ 1.5 billion to construct the Gerbi Dam in Ethiopia to provide water to Addis Ababa [27].

Investment in sanitation and water treatment infrastructure should be prioritised because water remains a critical link between agriculture and energy. Therefore, robust sanitation and water treatment infrastructure provides and supports opportunities in the agriculture, manufacturing and energy sectors but to mention a few [21]. The African agriculture sector offers and supports the highest number of jobs compared to any other sector [17]. Therefore, there is a need to assess and invest in agricultural water needs. The link between *Salmonella* infections and crop produce in HIC has been established [12]. This may make the export of African crop produce challenging based on the quality of water used for agricultural purposes. This justifies prioritising wastewater and sanitation infrastructure because of their positive impact on the quality of water used for agricultural purposes. Improving the quality of water used for agricultural purposes will enable the export of better-quality produce.

Investment in sanitation and water treatment infrastructure offers social and economic benefits. But why is it that African governments do not prioritise such critical infrastructure? Water infrastructure financing would require loans and hence a well-managed system of offering a paid water service [26]. Currently, the provision of paid water services remains a challenge. Thus, the servicing of water, wastewater, and sanitation infrastructure loans is associated with a high financial risk to the lender. Perhaps this is one of the reasons why social impact research and interventions have focused on point-of-use systems to ensure the availability of treated water for consumption at a household level. The only challenge with this approach is that the sanitation aspect may not receive the attention it deserves.

4. Water treatment at the household level

The chronic lack of sanitation and water treatment infrastructure in resource-poor communities, especially those in LMICs, makes the people living in these communities vulnerable to *Salmonella* infections. Point of Use (POU) water treatment systems have been suggested as a short to mid-term intervention to protect human health. The currently available POU water treatment systems work based on coagulation-flocculation, filtration, and disinfection [28].

4.1 Coagulation: Flocculation

Coagulation – flocculation-based systems offer a reliable, low, energy means of reducing the particulate matter in water, leaving it clearer than before. This approach would require using coagulants such as aluminium sulphate, lime, polyelectrolyte and iron salts (ferric chloride and ferric sulphate). Also, biopolymers, especially natural gums and bio-flocculants, have been investigated for their ability to serve as effective coagulants and flocculants [29]. However, the coagulation-flocculation treatments reduce turbidity the offer the added advantage of reducing the microbial burden of turbid water [29, 30]. Flocculation follows the addition of a coagulant. The flocculation process is often facilitated by gentle mixing to enable the formation of flocs. Mixing increases the collisions and interaction between the flocs and coagulant, thus increasing the size of the flocs resulting in them settling at the bottom by sedimentation. Coagulation – flocculation can be accessed at the POU using either the PUR or Poly Glu sachet manufactured by Procter & Gamble Co or Poly Glu International Co, respectively [28]. Both PUR and Poly are accessible worldwide. Nonetheless, extra measures are needed to ensure that these approaches are supplemented with either a disinfection method (PUR sachet) or proper hygiene handling of treated water to avoid recontamination (Poly Glu sachet) [28]. It should be noted that coagulation-flocculation-based POU solutions are often single-use and hence may be costly for some communities in the long run.

4.2 Filtration

Filtration systems offer a simple means of removing colloids, suspended solids, and microorganisms from water. Size exclusion is the basic principle behind the filtration process. As such, a properly configured filtration system can remove not only *Salmonella* or related bacteria from water but also viruses, toxins, and chemicals. This depends on the size of the pores on the filter membrane or biosand configuration. Membrane filtration systems used at the POU would ideally require one to consider the quality of the influent water and an external driving force relative to the membrane's pore size. These filters require maintenance because, with time, a foulant layer forms on them. This makes membrane-based filtration systems an option for communities that may have access to piped water that is not sufficiently treated. But inaccessible to those people in communities with no access to piped water.

Furthermore, membrane filtration is costly to maintain and would require technical skills to do so [28]. Sand-based filtration systems offer a more viable solution for those living in communities without access to treated piped water. Sand filters are easy to manufacture because the required raw materials are readily available. More than 500,000 people worldwide use biosand filters to meet their needs for potable water [28]. Biosand filters have been shown to reduce the turbidity of water. They have also been reported to reduce microbial contaminants but not to the level that meets the WHO water guidelines. Although the material to make biosand filters is easily accessible, the manufacturing process requires some technical skills. For instance, a correct balance between the flow rate and retention time is needed during manufacturing. These two variables have an inversely proportional relationship that influences the effectiveness of the removal of microbial contaminants such as *Salmonella* [28]. Also, the filter's depth needs to be considered to remove viruses.

4.3 Disinfection

Disinfection is an approach to enable the availability of safe water that relies on the destruction of the water contaminating microorganisms. Currently, two methods have been used to achieve disinfection: nanotechnology and Solar Disinfection (SODIS). Nanotechnology-based POU water treatment approach uses either titanium dioxide (TiO_2) or silver (Ag) nanoparticles for disinfection. The TiO_2 method requires a source of UV–vis light which facilitates the generation of hydroxyl radicals and hydrogen peroxide [31] that oxidise the organics, thus inactivating the microorganisms. TiO_2 is not depleted during this process, so the reaction continues. TiO_2 has been used to reduce biofouling on membranes used for water treatment and enhance the SODIS process [32]. TiO_2 has been used to develop a POU product, the Solarbag produced by Paralytics.

Ag nanoparticles are toxic to microorganisms. They bring about the death of microorganisms by either permeabilising the cell membranes or bioaccumulating causing irreversible damage to the DNA [33]. Ag nanoparticles have been used to coat ceramic [34] and polyurethane filters [35] to improve microbial log reduction. Currently, Ag nanoparticles are used in POU products such as Tata Swach and Folia filters to disinfect water. Although TiO_2 and Ag nanoparticles improve water microbiological quality, the long-term effects of these nanoparticles are not understood. At elevated levels, these nanoparticles are harmful to aquatic life [36, 37].

Furthermore, a high concentration of Ag nanoparticles has been shown to reduce mammalian cell vitality and mitochondrial function and cause cell membrane leakages [38]. This means that the use of nanoparticles to improve the quality of the water needs to be supplemented with proper disposal of damaged, unusable systems and accumulated waste. Besides the use of nanoparticles, natural sunlight could be used to sterilise the water before its consumption.

5. Solar disinfection of water

SODIS of water is an affordable and easy method of treating microbiologically contaminated water before its consumption. As such this section will focus on SODIS as opposed to the other approaches. During SODIS, microbiologically contaminated water in a transparent clear vessel is exposed to natural sunlight for approximately 6 hours on a sunny day with clear skies and on two rainy days overcast days. A detailed manual on the application of SODIS in the field has been developed and is accessible via the Swiss Federal Institute of Aquatic Science and Technology [39].

Effective bacterial inactivation is judged by the inability of the microorganisms to form colonies after SODIS treatment [40]. Downes and Blunt [41] were the first to present empirical evidence of the bactericidal effect of sunlight; however, its use to sanitise water can be traced as far back as 2000 BC. Presently, Downes and Blunts [41] observations on the bactericidal effect of natural sunshine are continuously confirmed by various research teams with consequent successful application in many countries globally. Studies by Acra et al. [42] and Conroy et al. [43] hypothesise that the observed bactericidal effect following sunshine exposure is due to the ultraviolet component of sunlight.

The harmful effects on the microbial population during SODIS are due to solar ultraviolet radiation (SUVR), which comprises wavelengths shorter than 400 nm. Natural sunlight reaching the earth's surface contains 6% of SUVR [44]. The UV wavelength is subdivided into three wavebands categorised as UVA (400–320 nm), UVB (320–280 nm) and UVC (280–100 nm) [45]. Of these three wavebands, UVA is the most abundant (95%) form of SUVR reaching the earth's surface, followed by UVB; UVC rarely reaches the earth's surface because the stratospheric ozone layer absorbs it. The amount of Solar Ultra-Violet Radiation (SUVR) reaching a given location on the earth's surface is influenced by geographical, meteorological and temporal factors such as the latitude, elevation, cloud cover, atmospheric conditions and ground reflection [45, 46]. The closer the exposure point is to the equator, the higher the levels of SUVR [46, 47]. However, due to the sun's elevation in the sky, 75% of the daily SUVR is received between 0900 and 1500H, irrespective of the exposure point [45].

5.1 Factors influencing solar disinfection of water

Although SODIS may seem like an ideal means of sanitising microbiologically contaminated water, it is influenced by several factors. One key factor to consider when using SODIS is the weather conditions. Cloud cover affects the amount of SUVR received on the earth's surface. It has been observed and reported that the amount of SUVR reaching the earth's surface is less when the sky is cloudy than in a cloudless sky. However, the enhancement of SODIS technology by incorporating compound parabolic concentrators could provide efficient inactivation within a short time during cloudy days [48]. In the absence of SUVR enhancers, it is advisable to establish guidelines on the duration required to achieve the necessary solar radiation intensity (500 W/m^2) [49].

Besides the weather conditions, water turbidity has a significant influence on SODIS. Turbid waters have been shown to reduce the efficacy of the SODIS process [50, 51] and thus protect microbes from inactivation. According to the recommendation by EAWAG, water turbidity higher than 30 Nephelometric turbidity units needs to be pretreated before SODIS treatment [52]. This could be achieved through filtration or simple settling. Turbidity can also be reduced by flocculation using minerals like Alum (potassium sulfate) and seeds of plants like Moringa oleifera. The ability of both these flocculants to clarify water before SODIS treatment has been tested and shown promising results [53]. However, consideration must be given to the fact that adding any form of pre-treatment step elongates the overall time required for disinfection and may have cost implications.

The amount of oxygen present in the water before SODIS significantly influences the outcome. Oxygen plays a key role in forming highly reactive forms of oxygen (oxygen free radicals and hydrogen peroxides) during solar irradiation. These reactive molecules react with cell structures and kill pathogens [54]. SODIS is more effective in water containing high oxygen levels [52]. Therefore, the guidelines recommend vigorously hand-shaking the vessel to dissolve oxygen in the water [52].

The material from which the vessel to be solar irradiated is made significantly influences the outcome of SODIS. Different types of transparent plastic materials made from either polyethene terephthalate (PET) or polyvinylchloride (PVC) are good transmitters of light in the UV-A and visible range of the solar spectrum [55, 56]. Transparent clear bottles such as empty soda and water bottles made from PET and PVC could be used for SODIS. There have been some concerns regarding the leaching of chemicals from the plastic bottles used for SODIS, but this threat is negligible [57, 58].

The temperature has been reported synergies with SUVR to enhance the SODIS water process [50]. Giannakis et al. [59] showed that SODIS carried at temperatures between 50 and 60°C increased inactivation efficiency. Several approaches to enhance the thermal rate of microbial inactivation have been investigated, and these include (i) circulating water over a black surface in an enclosed casing that was transparent to UV-A light [60], (ii) painting sections of the bottles with black paint, and (iii) using a solar collector attached to a double glass envelope container [61]—increasing the temperature past the optimum growth temperature results in the destabilisation of the core structures of most proteins through denaturation. Denatured proteins cannot carry out their critical biological tasks, and as a result, the death of the affected microorganism may result. The increase in the water temperature has been attributed to infrared radiation from the sun.

5.2 The effect of SUVR on biological systems

UV's bactericidal effect involves thermal and optical processes [62]. Exposure of biological systems to SUVR results in wavelength-dependent outcomes [47, 63]. The observed physical effects are based on the absorbing molecules' action spectrum [47]. An action spectrum can be defined as a plot showing the relative effectiveness of radiations of different wavelengths to produce a given biological effect [47]. Therefore, the action spectrum leading to the formation of a particular photoproduct would be similar to the absorption spectrum of the molecules responsible for forming that photoproduct [47]. The damaging effects of SUVR on microorganisms are demonstrated by reduced exoenzymatic activity that often results in reduced DNA and protein synthesis, reduced amino acid uptake, reduced oxygen consumption and a decrease in bacterial abundance [64, 65]. Other biological entities, such as biofilms, greatly reduce the amount of SUVR absorption [66].

SUVR enables the formation of reactive oxygen species such as superoxide radicals, hydroxyl radicals, hydrogen peroxide and singlet oxygen. These reactive molecules, also known as photosensitisers, are formed through a process known as photo-oxidation [66–69]. During SODIS, the interaction between the photosensitisers and the actively growing microorganism results in irreversible damage to the microbial catalase systems rendering them susceptible to damage from peroxide formation [64, 70]. Furthermore, UVA, through photo-oxidation, blocks the electron transport chain (responsible for energy production), induces damage to the cell membrane, thus inactivating transport systems, and interferes with metabolic energy production, causing single-strand breaks in DNA [65, 71, 72]. Overall, UVA confers indirect multi-target damage to the microbial cellular components such as DNA, protein and lipids through the formation of photosensitisers [63].

Even though SUVR-exposed biological systems result in reduced functionality and destruction, protective cellular mechanisms are capable of reversing some of this damage. Several DNA repair mechanisms relevant to SUVR damage have been established, including photo reactivation repair, nucleotide excision repair (NER), post replication repair and SOS repair [47, 63, 73]. But these all depend on the dose of SUVR [53] and the exposure environment.

5.3 Solar disinfection of water an ideal POU

The efficacy of SODIS to inactivate a variety of pathogens such as *Vibrio cholera* [74], *Salmonella* Typhimurium [40], and *Shigella dysentriae* [40] has been demonstrated by various research teams. Millions of people in more than 50 countries, especially resource-poor communities, rely on SODIS-treated water [75, 76]. Input costs for low volume (< 5 litres) vessels are less than the other alternative approaches discussed in Section 4 above. Communities scale the process through the exposure of multiple vessels. Containers that can disinfect more than 5 litres significantly would require financial input. The Sustainable Sanitation and Water Management (SSWM) toolbox [44] offers a one stop hub for knowledge on SODIS where the SODIS manual [39] can also be accessed. The SSWM toolbox is an invaluable resource for organisations promoting access to clean water through the adoption of low-cost technologies such as SODIS.

6. Conclusion

Salmonella remains a critical pathogen of concern globally. This pathogen is responsible for the deaths of many children below the age of 5 and the fragile and elderly. Overcoming infections due to *Salmonella* would require that sanitary and water treatment infrastructure is prioritised, especially in LMIC. Resource-poor communities without access to sanitary or water treatment infrastructure could use a combination of coagulation-flocculation, filtration, and disinfection methods to access treated water at the POU. However, these methods do not address the frequent reintroduction of pathogens such as *Salmonella* into environmental waters. This requires the adoption of sanitary measures at a household level. The water treatment at the POU may reduce the burden of *Salmonella* transmitted through the consumption of contaminated water. However, *Salmonella* can be transmitted through the consumption of food of either animal or crop origin. Therefore, it is important to consider using some of these methods to treat agricultural water before its use. High water capacity SODIS interventions should be developed and evaluated for the provision of water for agricultural purposes. Perhaps this would require the combination of SODIS with other low-cost treatment approaches such as coagulation-flocculation.

Acknowledgements

I want to acknowledge the financial support provided by the Vaal University of Technology to support SODIS research.

Conflict of interest

The author declares no conflict of interest.

Author details

Cornelius Cano Ssemakalu
Faculty of Applied and Computer Sciences, Department of Biotechnology and Chemistry, Vaal University of Technology, Vanderbijlpark, South Africa

*Address all correspondence to: corneliuss@vut.ac.za

References

[1] Johnson R, Mylona E, Frankel G. TyphoidalSalmonella:Distinctivevirulence factors and pathogenesis. Cellular Microbiology. 2018;**20**:e12939

[2] Smith SI, Seriki A, Ajayi A. Typhoidal and non-typhoidal *Salmonella* infections in Africa. European Journal of Clinical Microbiology & Infectious Diseases. 2016;**35**:1913-1922

[3] Gal-Mor O, Boyle EC, Grassl GA. Same species, different diseases: How and why typhoidal and non-typhoidal Salmonella enterica serovars differ. Frontiers in Microbiology. 2014;**5**:10. DOI: 10.3389/FMICB.2014.00391 [Epub ahead of print]

[4] Uche IV, MacLennan CA, Saul A. A systematic review of the incidence, risk factors and case fatality rates of invasive nontyphoidal *Salmonella* (iNTS) disease in Africa (1966 to 2014). PLoS Neglected Tropical Diseases. 2017;**11**:e0005118

[5] Meiring JE, Shakya M, Khanam F, et al. Burden of enteric fever at three urban sites in Africa and Asia: A multicentre population-based study. The Lancet Global Health. 2021;**9**:e1688-e1696

[6] Lokken KL, Walker GT, Tsolis RM. Disseminated infections with antibiotic-resistant non-typhoidal *Salmonella* strains: Contributions of host and pathogen factors. Pathogens and Disease. 2016;**74**:103

[7] United Nations. *Salmonella* (non-typhoidal). Key Facts. Available from: https://www.who.int/news-room/fact-sheets/detail/salmonella-(non-typhoidal). 2018. accessed 9 June 2022

[8] Crump JA, Heyderman RS. A perspective on invasive *Salmonella* disease in Africa. Clinical Infectious Diseases : An Official Publication of the Infectious Diseases Society of America. 2015;**61**:S235

[9] Painter JA, Hoekstra RM, Ayers T, et al. Attribution of foodborne illnesses, hospitalizations, and deaths to food commodities by using outbreak data, United States, 1998-2008. Emerging Infectious Diseases. 2013;**19**:407-415

[10] World Health Organisation. Multi-country outbreak of *Salmonella Typhimurium* linked to chocolate products – Europe and the United States of America. Available from: https://www.who.int/emergencies/disease-outbreak-news/item/2022-DON369. 2022. accessed 2 June 2022

[11] Dekker D, Krumkamp R, Eibach D, et al. Characterization of *Salmonella* enterica from invasive bloodstream infections and water sources in rural Ghana. BMC Infectious Diseases. 2018;**18**:47

[12] Liu H, Whitehouse CA, Li B. Presence and persistence of *Salmonella* in water: The impact on microbial quality of water and food safety. Frontiers in Public Health. 2018;**6**:159

[13] Hu B, Hou P, Teng L, et al. Genomic investigation reveals a community typhoid outbreak caused by contaminated drinking water in China, 2016. Frontiers in Medicine. 2022;**9**:448

[14] Ekwanzala MD, Abia ALK, Keshri J, et al. Genetic characterization of *Salmonella* and *Shigella* spp. isolates recovered from water and riverbed sediment of the Apies River, South Africa. Water SA. 2017;**43**:387-397

[15] Carstens CK, Salazar JK, Darkoh C. Multistate outbreaks of foodborne illness

in the United States associated with fresh produce from 2010 to 2017. Frontiers in Microbiology. 2019;**10**:2667

[16] Berger CN, Sodha SV, Shaw RK, et al. Fresh fruit and vegetables as vehicles for the transmission of human pathogens. Environmental Microbiology. 2010;**12**:2385-2397

[17] Kamiński M, Skonieczna-Żydecka K, Nowak JK, et al. Global and local diet popularity rankings, their secular trends, and seasonal variation in Google Trends data. Nutrition. 2020;**79-80**:110759

[18] Kovačić A, Huljev Ž, Sušić E. Ground water as the source of an outbreak of *Salmonella Enteritidis*. Journal of Epidemiology and Global Health. 2017;7:181-184

[19] UN. The Sustainable Development Goals Report 2021. Available from: https://unstats.un.org/sdgs/report/2021/. 2021. accessed 9 June 2022

[20] OECD. Benefits of investing in water and sanitation: An OECD perspective, OECD studies on water. Paris. doi: 10.1787/9789264100817-en. 2011. accessed 17 June 2022

[21] IWA, IUCN, ICA. Nexus trade-offs and strategies for addressing the water, agriculture and energy security nexus in Africa. Geneva. Available from: https://www.icafrica.org/fileadmin/documents/Publications/Nexus_Trade-off_and_Strategies__ICA_Report__June2016_2_1_.pdf. 2015. accessed 17 June 2022

[22] United Nations Department of Economic and Social Affairs Population Division. World Population Prospects 2022: Summary of Results. New York. Available from: www.unpopulation.org. 2022. accessed 6 October 2022

[23] The World Bank. Population ages 0-14 (% of total population) - Sub-Saharan Africa, World, Middle East & North Africa, East Asia & Pacific, Europe & Central Asia, North America, Latin America & Caribbean, Australia — Data. Data. Available from: https://data.worldbank.org/indicator/SP.POP.0014.TO.ZS?end=2021&locations=ZG-1W-ZQ-Z4-Z7-XU-ZJ-AU&name_desc=false&start=2021&view=bar. accessed 6 October 2022

[24] Suzuki E. World's population will continue to grow and will reach nearly 10 billion by 2050. World Bank Blogs. Available from: https://blogs.worldbank.org/opendata/worlds-population-will-continue-grow-and-will-reach-nearly-10-billion-2050. 2019. accessed 6 October 2022

[25] Ezeh A, Kissling F, Singer P. Why sub-Saharan Africa might exceed its projected population size by 2100. Lancet. 2020;**396**:1131-1133

[26] Eberhard R. Access to water and sanitation in Sub-Saharan Africa. Stresemannstraße 94. Available from: www.giz.de. 2019. accessed 17 June 2022

[27] The Infrastructure Consortium for Africa, African Development Bank. Infrastructure Financing Trends in Africa – 2017. Abidjan 01. Available from: https://www.icafrica.org/fileadmin/documents/Annual_Reports/IFT2017.pdf. 2018. accessed 17 June 2022

[28] Pooi CK, Ng HY. Review of low-cost point-of-use water treatment systems for developing communities. npj Clean Water. 2018;**1**:1-8

[29] Evelyn Z-P, Neftalí R-V, Isaac C, et al. Coliforms and helminth eggs removals by coagulation-flocculation treatment based on natural polymers. Journal of Water Resource and Protection. 2013;**5**(11):1027-1036

[30] Okaiyeto K, Nwodo UU, Okoli SA, et al. Implications for public health demands alternatives to inorganic and synthetic flocculants: Bioflocculants as important candidates. Microbiology. 2016;**5**:177

[31] Kumar SG, Devi LG. Review on modified TiO2 photocatalysis under UV/visible light: Selected results and related mechanisms on interfacial charge carrier transfer dynamics. The Journal of Physical Chemistry. A. 2011;**115**:13211-13241

[32] Gelover S, Gómez LA, Reyes K, et al. A practical demonstration of water disinfection using TiO2 films and sunlight. Water Research. 2006;**40**:3274-3280

[33] Yin IX, Zhang J, Zhao IS, et al. The Antibacterial mechanism of silver nanoparticles and its application in dentistry. International Journal of Nanomedicine. 2020;**15**:2555

[34] Ngoc Dung TT, Phan Thi LA, Nam VN, et al. Preparation of silver nanoparticle-containing ceramic filter by in-situ reduction and application for water disinfection. Journal of Environmental Chemical Engineering. 2019;**7**:103176

[35] Jain P, Pradeep T. Potential of silver nanoparticle-coated polyurethane foam as an antibacterial water filter. Biotechnology and Bioengineering. 2005;**90**:59-63

[36] Abdel-Latif HMR, Dawood MAO, Menanteau-Ledouble S, et al. Environmental transformation of n-TiO2 in the aquatic systems and their ecotoxicity in bivalve mollusks: A systematic review. Ecotoxicology and Environmental Safety. 2020;**200**:110776

[37] Fabrega J, Luoma SN, Tyler CR, et al. Silver nanoparticles: Behaviour and effects in the aquatic environment. Environment International. 2011;**37**:517-531

[38] Ferdous Z, Nemmar A. Health impact of silver nanoparticles: A Review of the biodistribution and toxicity following various routes of exposure. International Journal of Molecular Sciences. 2020;**21**:2375. DOI: 10.3390/IJMS21072375

[39] Luzi S, Tobler M, Suter F, et al. SODIS manual: Guidance on solar water disinfection. Dübendorf: Eawag. Available from: https://www.sodis.ch/methode/anwendung/ausbildungsmaterial/dokumente_material/sodismanual_2016.pdf. 2016. accessed 7 October 2022

[40] Smith RJJ, Kehoe SCC, McGuigan KGG, et al. Effects of simulated solar disinfection of water on infectivity of *Salmonella typhimurium*. Letters in Applied Microbiology. 2000;**31**:284-288

[41] Downes A, Blunt TP. Researches on the effect of light upon bacteria and other organisms. Proceedings of the Royal Society of London. 1877;**26**:488-500

[42] Acra A, Jurdi M, Mu'Allem H, et al. Sunlight as disinfectant. Lancet. 1989;**1**:280

[43] Conroy RM, Elmore-Meegan M, Joyce T, et al. Solar disinfection of drinking water and diarrhoea in Maasai children: A controlled field trial. Lancet. 1996;**348**:1695-1697

[44] Dorothee S, Regula M. SODIS — SSWM - Find tools for sustainable sanitation and water management! Sustainable Sanitation and Water Management. Available from: https://sswm.info/sswm-solutions-bop-markets/affordable-wash-services-and-products/

affordable-water-supply/sodis. accessed 7 October 2022

[45] Parisi AV. Physics concepts of solar ultraviolet radiation by distance education. European Journal of Physics. 2005;**26**:313-320

[46] Diffey BL, Roscoe AH. Exposure to solar ultraviolet radiation in flight. Aviation, Space, and Environmental Medicine. 1990;**61**:1032-1035

[47] Diffey BL. Solar ultraviolet radiation effects on biological systems. Physics in Medicine and Biology. 1991;**36**:299-328

[48] Ubomba-Jaswa E, Fernandez-Ibanez P, Navntoft C, et al. Investigating the microbial inactivation efficiency of a 25 L batch solar disinfection (SODIS) reactor enhanced with a compound parabolic collector (CPC) for household use. Journal of Chemical Technology and Biotechnology. 2010;**85**:1028-1037

[49] Nwankwo EJ, Agunwamba JC, Nnaji CC. Effect of radiation intensity, water temperature and support-base materials on the inactivation efficiency of solar water disinfection (SODIS). Water Resources Management. 2019;**33**:4539-4551

[50] Dessie A, Alemayehu E, Mekonen S, et al. Solar disinfection: An approach for low-cost household water treatment technology in Southwestern Ethiopia. Journal of Environmental Health Science and Engineering. 2014;**12**:25

[51] Asiimwe JK, Quilty B, Muyanja CK, et al. Field comparison of solar water disinfection (SODIS) efficacy between glass and polyethylene terephalate (PET) plastic bottles under sub-Saharan weather conditions. Journal of Water and Health. 2013;**11**:729-737

[52] Eawag, Sandec. Solar Water Disinfection a guide for the application of Sodis. Duebendorf. Available from: https://ec.europa.eu/echo/files/evaluation/watsan2005/annex_files/SKAT/SKAT1 -Solar disinfection of water/Manual - solar disinfection of water - SODIS.pdf. October 2002. accessed 11 July 2022

[53] Asrafuzzaman M, Fakhruddin ANM, Hossain MA. Reduction of turbidity of water using locally available natural coagulants. ISRN Microbiology. 2011;**2011**:1-6

[54] Fisher MB, Nelson KL. Inactivation of *Escherichia coli* by polychromatic simulated sunlight: Evidence for and implications of a fenton mechanism involving iron, hydrogen peroxide, and superoxide. Applied and Environmental Microbiology. 2014;**80**:935-942

[55] Borde P, Elmusharaf K, McGuigan KG, et al. Community challenges when using large plastic bottles for solar energy disinfection of water (SODIS). BMC Public Health. 2016;**16**:1-8

[56] Johansson J, Aguirre Ramirez NJ, Escobar Tovar C, et al. Solar disinfection at low costs: An experimental approach towards up-scaled continuous flow systems. H2Open Journal. 2022;**5**:153-165

[57] Ozores Diez P, Giannakis S, Rodríguez-Chueca J, et al. Enhancing solar disinfection (SODIS) with the photo-Fenton or the Fe2+/peroxymonosulfate-activation process in large-scale plastic bottles leads to toxicologically safe drinking water. Water Research. 2020;**186**:116387

[58] Schmid P, Kohler M, Meierhofer R, et al. Does the reuse of PET bottles during solar water disinfection pose a health risk due to the migration of plasticisers and other chemicals

into the water? Water Research. 2008;**42**:5054-5060

[59] Giannakis S, Darakas E, Escalas-Cañellas A, et al. The antagonistic and synergistic effects of temperature during solar disinfection of synthetic secondary effluent. Journal of Photochemistry and Photobiology A: Chemistry. 2014;**280**:14-26

[60] Martín-Domínguez A, Alarcón-Herrera MT, Martín-Domínguez IR, et al. Efficiency in the disinfection of water for human consumption in rural communities using solar radiation. Solar Energy. 2005;**78**:31-40

[61] Saitoh TS, El-Ghetany HH. A pilot solar water disinfecting system: Performance analysis and testing. Solar Energy. 2002;**72**:261-269

[62] Conroy RM, Meegan ME, Joyce T, et al. Solar disinfection of water reduces diarrhoeal disease: An update. Archives of Disease in Childhood. 1999;**81**:337-338

[63] Joux F, Jeffrey WH, Lebaron P, et al. Marine bacterial isolates display diverse responses to UV-B radiation. Applied and Environmental Microbiology. 1999;**65**:3820-3827

[64] Alonso-Sáez L, Gasol JM, Lefort T, et al. Effect of natural sunlight on bacterial activity and differential sensitivity of natural bacterioplankton groups in Northwestern Mediterranean coastal waters. Applied and Environmental Microbiology. 2006;**72**:5806-5813

[65] Bosshard F, Riedel K, Schneider T, et al. Protein oxidation and aggregation in UVA-irradiated *Escherichia coli* cells as signs of accelerated cellular senescence. Environmental Microbiology. 2010;**12**:2931-2945

[66] Elasri MO, Miller RV. Study of the response of a biofilm bacterial community to UV radiation. Applied and Environmental Microbiology. 1999;**65**:2025-2031

[67] Navntoft C, Ubomba-Jaswa E, McGuigan KG, et al. Effectiveness of solar disinfection using batch reactors with non-imaging aluminium reflectors under real conditions: Natural well-water and solar light. Journal of Photochemistry and Photobiology B: Biology. 2008;**93**:155-161

[68] Qiu X, Sundin GW, Chai B, et al. Survival of *Shewanella* oneidensis MR-1 after UV radiation exposure. Applied and Environmental Microbiology. 2004;**70**:6435-6443

[69] Sinton LW, Finlay RK, Lynch PA. Sunlight inactivation of fecal bacteriophages and bacteria in sewage-polluted seawater. Applied and Environmental Microbiology. 1999;**65**:3605-3613

[70] Bailey CA, Neihof RA, Tabor PS. Inhibitory effect of solar radiation on amino acid uptake in *Chesapeake Bay* bacteria. Applied and Environmental Microbiology. 1983;**46**:44-49

[71] Berney M, Weilenmann HU, Simonetti A, et al. Efficacy of solar disinfection of *Escherichia coli*, *Shigella flexneri*, *Salmonella Typhimurium* and *Vibrio cholerae*. Journal of Applied Microbiology. 2006;**101**:828-836

[72] Bosshard F, Bucheli M, Meur Y, et al. The respiratory chain is the cell's Achilles' heel during UVA inactivation in *Escherichia coli*. Microbiology. 2010;**156**:2006-2015

[73] Arrage AA, Phelps TJ, Benoit RE, et al. Survival of subsurface microorganisms exposed to UV radiation

and hydrogen peroxide. Applied and Environmental Microbiology. 1993;**59**:3545-3550

[74] Ssemakalu CC. The effect of solar ultraviolet radiation and ambient temperature on the culturability of toxigenic and non-toxigenic Vibrio cholerae in Pretoria, South Africa. African Journal of Microbiology Research. 2012;**6**:5957-5964

[75] Moreno-SanSegundo J, Giannakis S, Samoili S, et al. SODIS potential: A novel parameter to assess the suitability of solar water disinfection worldwide. Chemical Engineering Journal. 2021;**419**:129889

[76] Meierhofer R, Landolt G. Factors supporting the sustained use of solar water disinfection — Experiences from a global promotion and dissemination programme. Desalination. 2009;**248**:144-151

Chapter 6

Biocide Use for the Control of Non-Typhoidal *Salmonella* in the Food-Producing Animal Scenario: A Primary Food Production to Fork Perspective

João Bettencourt Cota, Madalena Vieira-Pinto and Manuela Oliveira

Abstract

Biocides are a group of substances commonly used in food production settings to destroy or control a wide range of microorganisms, which can be present in food of animal origin, since contamination can occur in the several steps of the food production chains. In order to achieve the desired results, the users of biocides must first understand the diverse characteristics of such compounds, mainly the usage requirements, limitations, and the factors affecting the activity of biocides. Food-producing animals and their products, namely meat and eggs, represent a major source of non-typhoidal *Salmonella* for humans and are associated with foodborne outbreaks worldwide. The prevention of cross-contamination, which can occur in any step of the food production chain, is essential for the ultimate objective of producing safe food products. The correct use of biocides, along with good hygiene and manufacturing practices, is one of the pillars of *Salmonella* spp. control and should be implemented in all steps of the food production chain. The present chapter reviews the accumulated knowledge on the use of biocides to control non-typhoidal *Salmonella*, from a farm to fork standpoint, along with the possible impacts on human health arising from improper use.

Keywords: biocides, non-typhoidal *Salmonella*, control, farm to fork, food safety, food production chain

1. Introduction

Biocides, from a broad point of view, are substances with the ability of killing living organisms, meaning that this is an all-embracing group, which includes numerous active substances with different targets, ranging from animals, plants, to microorganisms. The use of biocides specifically targeting microorganisms is widely spread in modern societies, mainly due to an increased alarm regarding microbial environmental contamination of living spaces [1]. Regardless of the growing usage of such biocides, antimicrobial chemical substances have long been regarded as very useful for mankind, for medical, agricultural, and food safety purposes [2]. Unlike antibiotics, which are used to treat infections in humans and animals since they are suitable to be in contact with living tissues, antimicrobial biocides are applied on contaminated suspensions or surfaces reducing the numbers or eliminating micro-organisms [1]. These substances are available in very diverse formulations and used not only at an industrial level, but also at the households of consumers, for multiple sanitation procedures. Likewise, these biocidal substances are also used to control the dissemination of microbial pathogens among animal populations and to prevent the leakage of such pathogens from farms [2]. The selection of the most appropriate antimicrobial biocide for a specific application is highly dependent on multiple factors, which can seriously affect its effectiveness [3]. Even with the growing concern regarding the possible effects of such a vast use of these substances in various sectors, antimicrobial biocides are considered to be indispensable for food safety assurance, as their use is imperative along the food production chains, from livestock production up to food industries and retailers [4].

Non-typhoidal *Salmonella* (NTS) is one of the most notorious and studied food-borne pathogens worldwide due to its impact on human health, with an estimated burden of 93.8 million cases of disease and 155.00 deaths per year globally, affecting populations of both developing and developed countries [5]. In humans, NTS infection cases are commonly restricted to a self-limiting gastroenteritis, characterized by nausea, vomiting, and diarrhea starting within a 6–48 hours interval after exposure; however, life-threatening complications can arise from the initial gastrointestinal tract infection in more susceptible groups, such as infants or immunosuppressed and HIV-positive individuals, among others [6, 7]. Despite not being considered necessary for uncomplicated human infections, empirical antimicrobial therapy should be considered in patients belonging to the increased risk groups and recommended whenever bloody diarrhea is present [8]. The upsurge of antimicrobial resistant NTS isolates seen over the past decades is therefore worrying, and this phenomenon has long been identified as a serious global public health concern [9]. As mentioned, NTS is generally considered to be a foodborne pathogen, though human infection cases can occur without the ingestion of contaminated food [6]. Nevertheless, the epidemiological role of food in NTS outbreaks is strikingly greater when comparing with other sources of infection, as direct animal contact or with animal environments [10, 11]. Additionally, food of animal origin has been largely implicated in NTS foodborne outbreaks when comparing with produce [12–14]. The major food vehicles of animal origin associated with outbreaks over the years have been eggs, poultry meat, pork, and to lesser extent, beef and dairy products [15]. Previous works have highlighted the public health impact of eggs [16], poultry and poultry meat [17], and pork [18, 19] in the salmonellosis scenario. There are several steps along the food production chains in which NTS can unintentionally taint food; therefore, complex strategies to avoid the presence of this foodborne pathogen in the final product must be adopted.

This chapter aims to provide a straightforward review of the most relevant available information regarding the use of antimicrobial biocides for the control of non-typhoidal *Salmonella* in the multiple points of the animal-origin food chains, and its possible implications, with a farm to fork perspective. A brief description concerning antimicrobial biocides and their main characteristics will be presented. Additionally, information regarding non-typhoidal *Salmonella* and its dissemination along the food chains will be reviewed. Finally, the use of biocides to control non-typhoidal *Salmonella*, biocide resistance, and possible implications of biocide usage will be discussed.

2. Biocides

Generally, a biocide can be defined as an active substance, or a formulation containing at least one active substance, used with the intention of destroying or controlling the effect of any harmful organism to human or animal health by any means other than mere physical or mechanical action [20]. Since the term biocide encompasses a wide spectrum of substances with diverse applications, in the scientific literature it is common to be replaced by disinfectant or sanitizer when addressing chemical substances with antimicrobial activity, in part due to different classifications and legislations. Within the scope of this chapter, only biocides used mainly for disinfection purposes will be addressed.

The legislation and the agencies that regulate these chemical substances have suffered changes over passed decades, mainly in the European Union (EU) and in the United States of America (USA). According to the EU's legislation, biocides are divided in four main groups regarding their purpose: disinfectants, preservatives, pest control products, and other biocidal products [20]. The EU's Biocidal products regulation (Regulation (EU) No 528/2012) further divides biocides used for disinfection in five groups: human hygiene biocidal products, private area and public health area disinfectants, veterinary hygiene biocidal products, food and feed area disinfectants, and drinking water disinfectants.

A different classification is seen in the USA as biocides with antimicrobial activity are classified as public health antimicrobial pesticides and are under the authority of the Federal Insecticide, Fungicide and Rodenticide Act (FIFRA). Within the US legislation, these antimicrobial pesticides are classified according to the degree of effectiveness as sterilants, disinfectants, and sanitizers. While sterilants destroy all forms of bacteria and fungi, including their spores, and even viruses, disinfectants destroy or irreversibly inactivate bacteria, fungi, and/or viruses but not their spores. Disinfectants are subdivided based on their efficacy as hospital, general or broad-spectrum, and limited disinfectants. With the lowest efficacy of all the public health antimicrobial pesticides, sanitizers reduce, without necessarily eliminating microorganisms from inanimate environment, and are divided as non-food-contact sanitizers and food-contact sanitizers [21].

2.1 Antimicrobial biocides

In terms of disinfection purposes, there are several biocidal active substances deriving from different chemical categories [22]. Overall, disinfectants can basically be divided into two groups, the oxidizing and the nonoxidizing. Among oxidizing disinfectants are halogens such as chlorine, chlorine dioxide, iodine, and peroxides, mostly peracetic acid and hydrogen peroxide. Within the group of nonoxidizing disinfectants are quaternary ammonium compounds (QAC), amphoterics, aldehydes, phenolic compounds, biguanides, and acid anionic agents [23, 24]. Their activity, and ultimately the desired effect, can be influenced by different factors, mainly the initial concentration, length of time of contact, temperature, pH, the presence of organic matter, and the type of surface [25–27]. Together with external factors, the nature of the microorganisms, their number, location, and condition, namely the presence of a biofilm, can also have an impact on the activity of biocides [27, 28]. When these factors are not considered, ineffective disinfection procedures are likely to occur [29]. The typical usage of antimicrobial biocides, the factors affecting their activity, their advantages and disadvantages have been summarized by different authors in previously published reviews [3, 29, 30].

The mechanisms of action of biocides are not fully understood, but generally they can be divided according to the cell structures in the bacterial cells where the interactions occur to produce an antimicrobial effect, specifically the outer cell components, the cytoplasmic membrane, or the cytoplasmic components [31]. In order to develop their antimicrobial activity, the biocidal substance must be transported to the bacterial cell surface, adsorb, diffuse, penetrate, and interact with its target, and all of these processes are time-dependent [32]. In fact, after biocide exposure, the bacterium expresses multiple mechanisms to reduce the amounts of biocidal substance and to repair damages. Consequently, if the exposure is short, the stress and damage induced by the biocide are reversible, but long exposures lead to cell death due to irreversible changes in membrane integrity, leakage of cytoplasmic constituents, and coagulation of intracellular materials [2].

Despite being used for the same reasons and aiming for similar outcomes, some of the characteristics of the biocides used in animal production settings are different when comparing to the ones used in food processing environments. Biocides used for disinfection of animal houses are usually strong, and on some occasions, such as contaminated surfaces, toxic biocidal chemicals are used; in contrast, biocides used in food processing premises are commonly of low toxicity and applied in higher dilutions [26].

Though precise information regarding the actual biocidal substances being used on farms is not readily available since there are several commercially available disinfectant formulations, among the most common are hydrogen peroxide, acetic acid, QACs, aldehydes such as glutaraldehyde, formaldehyde, and isopropanol [33]. In the food industry, the biocidal substances used in commercially available formulations include amphoteric surfactants, polymeric biguanides, QACs, chlorhexidine, chlorine and chlorine-based derivatives, acid anionic agents, hydrogen peroxide, and peracetic acid since these biocidal groups are suitable to be used on food-contact surfaces [4, 32].

These substances or products are extremely important and broadly used for cleaning and disinfection (C&D) procedures of surfaces and environments in the multiple steps of the food production chain, from farms to abattoirs and food processing and handling establishments and even at the households of consumers [30, 34]. As previously mentioned, NTS is a major foodborne illness hazard, thus controlling its movement and persistence across the food production chains is imperative to diminish its impact on human health.

3. Non-typhoidal *Salmonella*

Despite belonging to the same species (i.e., *Salmonella enterica*), non-typhoidal and typhoidal *Salmonella* serotypes have very distinct behaviors regarding the hosts. While typhoidal *Salmonella* serotypes, specifically Typhi and Paratyphi, are highly adapted to the human host, NTS serotypes can infect a broad range of hosts, including humans, though some NTS serotypes are also known to be species restricted [35]. This level of adaptation of each serotype to specific hosts has clinical, epidemiological, and public health impacts, since the degree of pathogenicity of the same serotype can vary among different hosts. As previously mentioned, *Salmonella* Typhi and Paratyphi, which are highly adapted serotypes to humans and are the etiological agents of typhoid and paratyphoid fevers, respectively, are not considered to be pathogenic to other animals. A similar scenario is observed regarding serotypes highly adapted to animal hosts, namely *Salmonella* Gallinarum responsible for fowl typhoid, which is not considered to be pathogenic to humans. On the other hand, ubiquitous or generalist serotypes, such as *Salmonella* Enteritidis or Typhimurium, can affect a broad range of hosts, including humans [36] and are among the most frequently implicated in NTS-associated foodborne illness cases [37, 38]. It is assumed that infections with generalist serotypes are mainly characterized by gastrointestinal manifestations, with high morbidity but with low mortality, and that diseases arising from host-restricted serotypes have low morbidity and high mortality [39]. Nevertheless, some exceptions to this host adaption/pathogenicity degree association are known to occur, for example, *Salmonella* Choleraesuis and Dublin, two serotypes that have as primary hosts pigs and cattle, respectively, which are also responsible for systemic disease in humans [36]. Within the scope of the present chapter, the use of NTS will be replaced simply by *Salmonella*.

3.1 Food production chains and *Salmonella*

The food production chains have evolved greatly since the past century. The world's most industrialized countries have seen a paradigm change on how food is produced, shifting from small-sized farms supplying local markets to international networks producing and supplying food to large amounts of consumers, though it is estimated that 50–70% of the global food is still produced by smallholder farmers [40]. With a projected world population of almost 10 billion by 2050, and an expected growth of the income in low and middle-income countries, a higher consumption of meat, fruits, and vegetables is foreseen, resulting in additional efforts in the production chains and on natural resources [41]. These circumstances highlight the global challenge of producing enough food to satisfy the needs of the world's growing population, but in order to do so, food safety systems will also have to adapt to the changing needs of both developed and developing countries, enabling global food security [42].

Many stakeholders take part in the food of animal-origin production chains, ranging from cereal producers, feed mills, animal farms, transport operators, abattoirs to food processing industries. These networks of stakeholders can be extremely intricate and highly dependent of international trade, with globalization having a very important role. Feed ingredients can, in some cases, originate from different continents, traveling long distances before being processed in feed mills. The role of feed as a source of *Salmonella* for animals and humans is well known, and all efforts should be made to avoid feed contamination. In the first place, it involves preventing the entry of *Salmonella* in the feed mill's facilities by obtaining uncontaminated feed ingredients and managing several other factors, including flow of personnel and the control of unwanted animals (rodents and wild birds), among others [43].

When comparing different animal species, namely poultry and pigs, some variations in the production cycles are found, with a stratified organization of animal farms, such as breeder, multiplier and finishing or fattening farms, and as such live animal transport is necessary within and between countries. In fact, one of the main challenges regarding the control of *Salmonella* is the prevalence levels among animal populations. In Europe, several countries have implemented strict *Salmonella* surveillance and control programs for poultry (broilers, turkeys, and laying hens) [44–46] and, to a lesser extent, for pigs [47] and cattle [48–51]. Generally, these programs rely on the collection of samples for *Salmonella* detection and on the implementation of restrictions on farms whenever positive results are found. Additionally, a big emphasis is put on the application of biosecurity measures in farms as an effort to avoid the entry of *Salmonella*. Some of the most relevant biosecurity measures are associated with correct cleaning and disinfection (C&D) procedures of the houses where animals are reared in and of the transport vehicles [52, 53]. Moreover, each step of the life cycle of a food-producing animal (birth, rearing, slaughtering) can take place in a different region of the same country or even in different countries.

Finally, before being available to consumers, food-animal products must be carried to food processing facilities and/or to retailers where cross-contamination can occur. As reviewed by Carrasco et al. (2012), there are multiple scenarios where *Salmonella* can contaminate food through food handlers, food-contact surfaces, equipment, and utensils emphasizing the importance of preventive control measures, namely adequate sanitation procedures in food processing and handling facilities but also the consumer's knowledge on good hygiene practices [54].

There has been an increase of the number of food business operators adopting the vertical integration structure, connecting its upstream suppliers with the downstream buyers. The ultimate goal of integrative growth is to increase the business profitability by controlling the most important related activities [55]. Vertical integration is also considered to be a part of the food business operator's private control strategies to tackle food safety hazards along with Hazard Analysis and Critical Control Point (HACCP) systems and third-party certifications [56]. On the other hand, non-integrated food business operators are more likely to be affected by both upstream and downstream operators, not only regarding safety issues but also economically since they are more dependent.

The poultry industry, specifically the broiler sector, was the first to adopt a vertically integrated organization after World War II, during the 1950s, in the USA. Vertical integration of the pig sector was only achieved much later, due to technical and husbandry issues [57]. Nevertheless, at the present time these are the two main animal species reared by large vertically integrated food business operators, especially in high income countries.

Eggs, poultry meat, and pork are the main sources of human salmonellosis cases through contaminated food, and as such, stronger efforts to control *Salmonella* must be put in place along the poultry and pig-associated food production chains, namely the correct use of antimicrobial biocides.

4. Biocide use throughout the food production chain

To control the spread of *Salmonella* along the food production chain, several measures must be put in place at different stages starting at feed mills to assure high food safety standards. An efficient control of *Salmonella* in feed mills is based on blocking the entry of this pathogen firstly, reducing the chances of *Salmonella* multiplication within the facilities, and by rendering the final product *Salmonella*-free by using thermal process or adding chemicals to feed [43].

Despite the low-moisture environment found in feed mill facilities, which impairs bacterial multiplication, *Salmonella* persistence in such circumstances is known to occur, and it is associated with biofilm-forming capability [58]. In these situations, chemical disinfection is necessary to eliminate this source of feed contamination. Despite being a crucial step of the C&D procedure, it seems that physical cleaning can also contribute for the dissemination of the bacterial contamination within the mill facilities [59].

The use of disinfectant formulations combining aldehydes, namely formaldehyde and glutaraldehyde and QACs, applied at high concentrations has been pointed out as the most appropriate against *Salmonella* on surfaces that are not easily cleaned [60]. A direct application of a 30% formaldehyde commercial solution is able to reduce *Salmonella* contamination down to undetectable levels in different types of surfaces, including stainless steel, plastic, polypropylene haul bags, rubber belts, and rubber tires [61]. However, a 70% ethanol-based disinfectant (P3- AlcoDes) and a peroxygen-based disinfectant (Virkon S) were reported to be the most effective when used on surfaces outperforming other disinfectants, even those with a QAC-aldehyde formulation, under laboratory conditions [62].

The specificities of feed mills must be considered by the business operators when choosing the biocidal formulations to be used for disinfection, specifically the need to maintain low levels of moisture. Once detected, *Salmonella* contaminations must be dealt with as soon as possible and rigorous monitoring after C&D should provide information regarding the effectiveness of the procedure. When comparing the legislation of different countries, the responsibility is placed upon the business operators as they must assure the production of safe compound feed. Besides, the economic costs of implementing controls to obtain *Salmonella*-free feed are considered to be limited and that the prevention of dissemination of this pathogen to animals through feed is economically achievable, supporting the implementation of *Salmonella*-negative regulation [63].

The environments of the houses/farms where animals are raised in pose serious challenges when considering C&D procedures, mostly due to the amount of organic matter, construction materials used, and multiple fixtures. To obtain the best results possible, all animals should be moved out of the areas or houses before C&D can be started and new animals should only be moved in after C&D has been completed, a system commonly referred to as all in/all out.

There are multiple reports on the efficacy of C&D procedures for *Salmonella* control in poultry farms based on the application of different biocides, either from broiler [64–69], laying hen [70–73], or duck farms [74]. The most frequently used disinfectants were phenol-based, namely formaldehyde, glutaraldehyde, and QACs. Though the use of such substances is considered to result in effective C&D, the application of glutaraldehyde, formaldehyde, and peroxygen solutions at a concentration of 1% was unable to eliminate *Salmonella from a poultry house under experimental conditions* [75]. Wall and floor crevices, drinkers, feeders, and vents can be problematic since these areas/fixtures can promote bacterial persistence, mainly due to the accumulation of dust or organic matter protecting bacteria from the action of biocides [68, 69]. Incomplete disinfection of the houses or of the equipment, leading to *Salmonella* persistence, is likely to promote early *Salmonella* exposure to new laying hen flocks [71] and is considered to be one of the risk factors for the *Salmonella* status of broiler flocks at the end of the production cycle [67].

There are different types of disinfectant formulations, based on QACs, aldehydes, peroxygen or peracetic acid-based, iodine-based compounds or chlorocresols are available to be used on pig holdings for *Salmonella* control, though with diverse effectiveness levels [76]. Disinfectants based on sodium hypochlorite or QACs are believed to be able to eliminate *Salmonella* from pig houses when properly applied after a correct cleaning step [77]. Additionally, in pig housing settings, it seems that better results are achieved using concentrated phenolic disinfectants rather than peroxygen-based products [78]. Even though formulations using combinations of glutaraldehyde and QACs are more effective than iodine-based disinfectants, over-dilution of glutaraldehyde-QACs disinfectants affects its performance, leading to procedure failure and to *Salmonella* persistence in pig houses after C&D [76]. In pigs, as well as in poultry, the maintenance of *Salmonella* on the environment hinders the effects of all other biosecurity measures, such as feed or rodent control. The environment can be contaminated even though it looks clean or undergoes multiple C&D routines, contributing greatly for the transmission of *Salmonella* within pig farms [79].

Abattoirs are a paramount step for *Salmonella* cross-contamination control. Apparently healthy animals can be *Salmonella* carriers, which can easily contaminate the abattoir's facilities and/or equipment, transferring *Salmonella* to, or even infecting negative animals in the lairage area or transferring the pathogen to carcasses during the slaughtering processes. Due to the likely event of environment contamination, highly effective C&D procedures must be adopted. Disinfection in abattoirs can be carried out using one or more of the many formulations suitable to be used in the food industry premises including alcohols, chlorine-based compounds, QACs, oxidizing agents, persulfates, surfactants, and iodophors [80]. As an additional effort to reduce to possibility of cross-contamination, logistic slaughter should be implemented whenever the *Salmonella* status of the animals is known, meaning that *Salmonella*-positive animals should only be slaughtered after negative animals. The effectiveness of this measure is strictly dependent of the absence of *Salmonella* from the environment and equipment of the abattoir [81].

In pig slaughterhouses, it has been shown that a main source of carcass contamination is the lairage environment rather than the gut or the lymph nodes of the slaughtered animals [82]. When comparing different protocols for *Salmonella* elimination in lairage pens, a procedure combining the use of detergent, followed by a chlorocresol-based disinfectant and a final drying step of 24 h, was the most effective [83]. Though not suitable for food-contact surfaces, chlorocresol can be used in lairage pens in abattoirs as these areas only receive live animals. *Salmonella*-free lairage pens are extremely important to reduce cross-contamination in the beginning of the process; nevertheless, the following steps also have a significant impact on the carcass hygiene. While some slaughtering processes can reduce *Salmonella* carcass contamination, namely scalding and singeing, others can promote carcass contamination, including inefficient scalding, dehairing, polishing, evisceration, and dressing activities [84]. Accordingly, not only should there be good hygiene and manufacturing practices during slaughter and carcass preparation, but also a special attention should be given to C&D of the slaughter line equipment avoiding the possibility of *Salmonella* biofilm formation and environmental persistence.

As for pigs, the poultry slaughterhouses are a decisive step for *Salmonella* contamination. The poultry abattoir scenario has some major differences when comparing with pigs: the animals are moved in crates or cages, and they are not placed in pens before slaughtering, also the slaughter line is almost entirely mechanized and the slaughtering procedures are automated allowing to process, in some broiler abattoirs, up to 15.000 birds per hour. The transport crates and the slaughter equipment have been pointed as possible sources for *Salmonella* contamination [85, 86]. Poultry should only be transported from the farms to abattoirs in clean and disinfected crates. Though C&D reduces the numbers of *Salmonella* present in crates, persistence can be due to the presence of biofilms, improper application of biocides, recontamination, or even cross-contamination [87]. The slaughtering process of poultry encompasses different mechanized steps in intricate equipment, namely scalding, defeathering, evisceration, and chilling, which can ultimately increase the chances of *Salmonella* contamination [87]. The use of standard C&D protocols can in some cases fail to fully eliminate equipment contamination, namely from the plucking machine, after slaughtering *Salmonella*-positive flocks leading to the cross-contamination of *Salmonella*-free flocks slaughtered afterward [81].

Food safety is, and should always be, a top priority issue for food processing industries. Good hygiene and manufacturing practices along with a HACCP plan are essential for obtaining safe animal products. In order to maintain bacterial contamination levels, including *Salmonella*, in the working areas as low as possible C&D must be carried out routinely and effectively. The most relevant biocidal compounds used in the food industry are halogens, peroxygens, acids, and QACs [88]. Regarding egg packing centers, Wales, Taylor, and Davies have recently provided a review on the disinfectants allowed to be used on those facilities, namely QACs, amphoteric surfactants, non-ionic surfactants, sodium hypochlorite, and ancillary agents [89].

The persistence of *Salmonella* in food processing environments, mostly due to biofilm formation, specifically in food-contact surfaces and equipment, after C&D can be associated with insufficient procedures [88, 90]. Additionally, *Salmonella* biofilms in food processing facilities can be a serious problem as biofilms formed in food-contact surfaces can turn out to be a continuous source of food contamination [91]. Despite the multiple reports available on the efficacy of different biocidal substances or formulations on *Salmonella* biofilms under laboratory conditions, studies focusing on the application of such biofilm treatments on food processing facilities are lacking.

Though not applicable in the EU, some countries allow the use of biocides on raw meat/carcasses for decontamination purposes, some examples are provided. In the USA, the use of sodium hypochlorite, peroxyacetic acid, cetylpyridinium chloride, trisodium phosphate, among others, during immersion chilling is preconized for antimicrobial treatment of poultry carcasses [92]. For pig carcasses, the possibility of chemical decontamination seems to be mainly limited to the use of organic acids, namely acetic and lactic acid [93].

The increase of the application of antimicrobial biocides along the food chain was mainly impelled on the one hand by the implementation of stricter food safety regulations and on the other by consumers' requirements. The possible impacts of such a change are still being studied, but some of the unintentional side effects are already clear.

5. Possible implications of antimicrobial biocide usage

As with any other biologically active substance, the application of antimicrobial biocides in multiple settings raises concerns due to the possible implications on human, animal, or environmental health. Subsequently, there are legal requirements enforcing an environmental impact assessment and an authorization by the competent authorities before issuing a license for marketing new biocides or biocide formulations [3, 32]. Nevertheless, the usage of antimicrobial biocides is not deprived of risks, namely their toxicity to humans or the tendency to allow the establishment of biocide resistance [94]. Some antimicrobial biocides can be highly reactive with other substances or can produce direct toxic effects or sensitization on users after dermal or respiratory exposure [95–97]. Additionally, as part of their mechanism of action, these are non-selective compounds and thus can affect multiple organisms other than the intended but can also remain active in the environment after use since they are not easily biodegradable [25]. These characteristics are associated with the presence of biocides in aquatic ecosystems, posing an environmental threat [98]. Furthermore, the improper use of very aggressive antimicrobial biocides or the increase of their dosage to surpass resistance situations increases the possible negative impacts of biocide usage on public health [32]. In fact, the most commonly studied implication of antimicrobial usage is the upsurge of resistances either to antimicrobial biocides or cross-resistances with antibiotics. Any type of resistance to antimicrobial biocides or cross-resistances with antibiotics occurring in *Salmonella* must not be taken lightly, as these phenomena can hinder the previously effective C&D protocols and antibiotic therapeutics whenever necessary in severe salmonellosis cases in humans.

5.1 *Salmonella* resistance to biocides

The effectiveness of C&D protocols to eliminate or reduce *Salmonella* is mainly based on the antimicrobial activity of biocidal substances; thus, resistance to biocides can render the disinfection step useless. A brief overview regarding *Salmonella* antimicrobial biocide resistance is provided along with the possibility of antimicrobial resistance co-selection.

In the literature, multiple definitions for biocide resistance can be found, though perhaps the simplest definition is resistance occurs whenever bacteria survive after biocidal exposure in practical use [99]. The use of other terms such as reduced tolerance or reduced susceptibility as a synonym for resistance is also frequent and is based on increases of the minimum inhibitory concentrations or the minimum bactericidal concentrations, which are assessed under laboratory conditions, and such changes might not have any practical significance [2]. In fact, the bacteria ability to survive is not only dependent on the conditions in which the disinfectant is applied, namely concentration and physical state, but also on bacterial characteristics and on environmental settings [100]. As reviewed by Maillard (2018), after biocide exposure, the stress induced in bacteria leads to the expression of different mechanisms in an attempt to avoid irreversible damage and cell death. These mechanisms include the decrease of the concentration of the biocide in bacteria, either by reducing its penetration, by means of efflux pumps or enzymatic degradation, by physiological or metabolic changes or due to mutations [2].

Apart from the presence of the outer membrane with the lipopolysaccharide layer, characteristic of all Gram-negative bacteria, which acts as a blockade to the entry of unwanted substances, it seems that the major mechanisms for *Salmonella* biocide resistance rely on efflux and enzymatic degradation of biocides as well on mutations on biocide targets and overexpression of target proteins [101]. Among the various mechanisms, the AcrAB-TolC efflux system is the best studied in *Salmonella* and has been associated with resistance in different studies under controlled laboratory conditions [102–104]. Still, biocide-resistant *Salmonella* isolates recovered from field studies are thought to be uncommon [101].

Some of the most conclusive reports on *Salmonella* biocide resistance originating from livestock have been reviewed by Wales and Davies, focusing not only on resistance to numerous biocides but also on the possible co-selection of antibiotic resistance arising from biocide exposure [105]. It is assumed that biocide use can select antimicrobial resistant strains either by picking out biocide resistant bacteria with resistance determinants and mutations also responsible for antimicrobial resistance (cross-resistance) or by selecting bacteria with mobile genetic elements which encode several resistance determinants, simultaneously to biocides and antimicrobials (co-resistance) [34]. Despite the studies suggesting that such co-selection can occur [102, 106, 107], which can eventually have an impact in antimicrobial therapy, the conditions arranged in laboratories are supposed to be different from those observed in real-world practice and thus not accurate models to understand biocide interactions with bacteria in the environment [105]. The actual impact of biocide resistance is not fully understood, and it could be almost as important as antimicrobial resistance, making it a focus for future research [108].

6. Conclusions

The review presented has emphasized, in an uncomplicated manner, the usage of biocides to control *Salmonella* in the food of animal-origin production chains, mainly on poultry and pigs as the major sources, and the possible implications of using these antimicrobial biocides to control this foodborne pathogen, from feed to food or in other terms, from farm to fork.

The use of biocidal substances for disinfection purposes is critical for food safety purposes regarding the control of *Salmonella* along the complex food chains, which supply consumers nowadays. The correct implementation of C&D procedures must always take place in order to reduce the possibilities of *Salmonella* persistence in the environment, a major factor for cross-contamination. It is clear that, in most cases, failure to eliminate *Salmonella* is mainly associated with incorrect usage of biocides rather than a biocide resistance situation. The actual extent of biocide resistance in multiple bacterial pathogens from environmental and food samples should be studied, aiding for a rational usage of these substances or formulations. Nevertheless, with multiple biocidal formulations available in the market, there are several viable options to choose from, considering the different scenarios presented. Furthermore, the development of new biocide formulations, either based on phytochemicals or in nanoparticles ensuring an improved release of the antimicrobial active substances within the intricate structure of biofilms, seems to be promising. Whenever unsuccessful C&D is detected, all steps of the process must be revised, considering the possibilities of improper cleaning, human error on manipulation and application of the biocide, and finally, rotation of biocidal substances or formulations if needed.

Biocide use should not be looked as a panacea for *Salmonella*-associated food safety issues, but together with rigorous control and eradication programs at the herd level, good hygiene and manufacturing practices starting at feed mills up to the food processing industry, and even at the houses of consumers, the burden of salmonellosis in humans can be diminished. Likewise, the scientific community and the competent authorities should also raise the awareness of the consumers toward the possible impacts of the massive usage of household biocidal products as surrogates for good handling and hygiene practices.

This is a continuously growing field of knowledge to which multiple scientific areas are contributing. Further studies, both laboratory and field-based, are required so the most efficient, cost-effective, and safe disinfection protocols can be implemented in the several scenarios where they are irreplaceable.

Acknowledgements

This work was supported by CIISA – Centre for Interdisciplinary Research in Animal Health, Faculty of Veterinary Medicine, University of Lisbon, Project UIDB/00276/2020, and AL4Animals – Associate Laboratory for Animal and Veterinary Sciences Project LA/P/0059/2020 – AL4AnimalS (Funded by FCT – Fundação para a Ciência e Tecnologia IP).

Conflict of interest

The authors declare no conflict of interest.

Author details

João Bettencourt Cota[1*], Madalena Vieira-Pinto[2] and Manuela Oliveira[1]

1 Faculty of Veterinary Medicine, CIISA—Centre for Interdisciplinary Research in Animal Health, Associate Laboratory for Animal and Veterinary Sciences (AL4AnimalS), University of Lisbon, Lisbon, Portugal

2 Department of Veterinary Sciences, University of Trás-os-Montes and Alto Douro (UTAD), CECAV-Veterinary and Animal Research Centre, University of Trás-os-Montes and Alto Douro (UTAD), Associate Laboratory for Animal and Veterinary Sciences (AL4AnimalS), Vila Real, Portugal

*Address all correspondence to: joaobcota@fmv.ulisboa.pt

References

[1] White DG, McDermott PF. Biocides, drug resistance and microbial evolution. Current Opinion in Microbiology. 2001;**4**(3):313-317

[2] Maillard J-Y. Resistance of bacteria to biocides. Microbiological Spectroscopy. 2018;**6**(2):1-17

[3] Karsa DR. Biocides. Handb Cleaning/Decontamination Surfaces. 2007;**1**:593-623

[4] Donaghy JA, Jagadeesan B, Goodburn K, Grunwald L, Jensen ON, Jespers A, et al. Relationship of sanitizers, disinfectants, and cleaning agents with antimicrobial resistance. Journal of Food Protection. 2019;**82**(5):889-902

[5] Majowicz SE, Musto J, Scallan E, Angulo FJ, Kirk M, O'Brien SJ, et al. The global burden of nontyphoidal salmonella gastroenteritis. Clinical Infectious Diseases. 2010;**50**:882-889

[6] Acheson D, Hohmann EL. Nontyphoidal Salmonellosis. Clinical Infectious Diseases. 2001;**32**(2):263-269

[7] Crum-Cianflone NF. Salmonellosis and the gastrointestinal tract: More than just peanut butter. NIH Public Access. 2008;**2008**:424-431

[8] McDermott PF, Zhao S, Tate H. Antimicrobial resistance in nontyphoidal Salmonella. Microbiological Spectroscopy. 2018;**6**(4):1-26

[9] Su LH, Chiu CH, Chu C, Ou JT. Antimicrobial resistance in nontyphoid Salmonella serotypes: A global challenge. Clinical Infectious Diseases. 2004;**39**(4):546-551

[10] Marus JR, Magee MJ, Manikonda K, Nichols MC. Outbreaks of Salmonella enterica infections linked to animal contact: Demographic and outbreak characteristics and comparison to foodborne outbreaks—United States, 2009-2014. Zoonoses and Public Health. 2019;**66**(4):370-376

[11] Wikswo ME, Roberts V, Marsh Z, Manikonda K, Gleason B, Kambhampati A, et al. Enteric illness outbreaks reported through the national outbreak reporting system—United States, 2009-2019. Clinical Infectious Diseases. 2022;**74**(11):1906-1913

[12] Ford L, Moffatt CRM, Fearnley E, Miller M, Gregory J, Sloan-Gardner TS, et al. The epidemiology of Salmonella enterica Outbreaks in Australia, 2001-2016. Frontiers in Sustainable Food System. 2018;**2018**:2

[13] Snyder TR, Boktor SW, M'ikanatha NM. Salmonellosis outbreaks by food vehicle, serotype, season, and geographical location, United States, 1998 to 2015. Journal of Food Protection. 2019;**82**(7):1191-1199

[14] Schirone M, Visciano P. Trends of Major Foodborne Outbreaks in the European Union during the Years 2015-2019. Hygiene. 2021;**1**:106-119

[15] Pires SM, de Knegt L, Hald T. Estimation of the relative contribution of different food and animal sources to human Salmonella infections in the European Union. EFSA Support Publication. 2017;**8**(8):80

[16] Whiley H, Ross K. Salmonella and eggs: From production to plate. International Journal of Environmental Research and Public Health. 2015;**12**(3):2543

[17] Wessels K, Rip D, Gouws P. Salmonella in Chicken Meat: Consumption, outbreaks, characteristics, current control methods and the potential of bacteriophage use. Food. 2021;**10**(8):1742

[18] Bonardi S. Salmonella in the pork production chain and its impact on human health in the European Union. Epidemiology and Infection. 2017;**145**(8):1513-1526

[19] Campos J, Mourão J, Peixe L, Antunes P, Campos J, Mourão J, et al. Non-typhoidal Salmonella in the Pig Production Chain: A comprehensive analysis of its impact on human health. Pathogens. 2019;**8**(1):19

[20] European Parliament, European Council. Regulation (EU) No 528/2012 of the European Parliament and of the Council of 22 May 2012 concerning the making available on the market and use of biocidal products Text with EEA relevance. 2012.

[21] Hayes S. United States regulation of antimicrobial pesticides. Russell, Hugo Ayliffe's Princ Pract Disinfect Preserv Steriliz. 2012;**17**:269-276

[22] Gnanadhas DP, Marathe SA, Chakravortty D. Biocides – resistance, cross-resistance mechanisms and assessment.

[23] Fisher J. Types of disinfectant. In: Encyclopaedia of Food Science. Food Technology and Nutrition. Cambridge, USA: Massachusetts; 1993. pp. 1382-1385

[24] Sandle T. Disinfectants and biocides. Disinfection Decontamination Practised Handbook. 2018;7:33

[25] Michalak I, Chojnacka K. Biocides. Encyclopedia Toxicology. 2014;**1**:461-463

[26] Meade E, Slattery MA, Garvey M. Biocidal resistance in clinically relevant microbial species: A major public health risk. Pathogens. 2021;**10**(5):598

[27] Maillard JY, McDonnell G. Selection and use of disinfectants. In Practice. 2012;**34**(5):292-299

[28] Russell AD. Biocide use and antibiotic resistance: The relevance of laboratory findings to clinical and environmental situations. The Lancet Infectious Diseases. 2003;**3**(12):794-803

[29] Maillard JY. Testing the Effectiveness of Disinfectants and Sanitizers. In: Handbook of Hygiene Control in the Food Industry. Second ed. Cambridge, UK: Elsevier; 2016. pp. 569-586

[30] Jones IA, Joshi LT. Biocide use in the antimicrobial era: A review. Molecules. 2021;**26**(8):2276

[31] Maillard JY. Bacterial target sites for biocide action. Journal of Applied Microbiology. 2002;**92**(1):16S-27S

[32] Ribeiro M, Simões LC, Simões M. Biocides. Encycl Microbiol. 2019;**1**:478-490

[33] Committee on Emerging S, Identified Health Risks N. Assessment of the Antibiotic Resistance Effects of Biocides Antibiotic Resistance Effects of Biocides 2 About the Scientific Committees. 2009

[34] Ruiz L, Alvarez-Ordóñez A. The role of the food chain in the spread of antimicrobial resistance (AMR). Functional Nanomaterials. 2017;**1**:23-47

[35] Gal-Mor O, Boyle EC, Grassl GA. Same species, different diseases: How and why typhoidal and non-typhoidal Salmonella enterica serovars differ. Frontiers in Microbiology. 2014;**2014**:5

[36] Chen HM, Wang Y, Su LH, Chiu CH. Nontyphoid Salmonella infection: Microbiology, clinical features, and antimicrobial therapy. Pediatrics and Neonatology. 2013;**54**:147-152

[37] Food Safety Authority E, Boelaert F, Amore G, Messens W, Hempen M, Rizzi V, et al. The European Union One Health 2020 Zoonoses Report. EFSA Journal. 2021;**19**(12):e06971

[38] Brown AC, Grass JE, Richardson LC, Nisler AL, Bicknese AS, Gould LH. Antimicrobial resistance in Salmonella that caused foodborne disease outbreaks: United States, 2003-2012. Epidemiology and Infection. 2017;**145**(4):766-774

[39] Hoelzer K, Switt AIM, Wiedmann M. Animal contact as a source of human non-typhoidal salmonellosis. Veterinary Research. 2011;**42**:39

[40] Giller KE, Delaune T, Silva JV, Descheemaeker K, van de Ven G, Schut AGT, et al. The future of farming: Who will produce our food? Food Security. 2021;**13**(5):1073-1099

[41] Food and Agriculture Organization of the United Nations. The Future of Food and Agriculture. Food Agric Organ United Nations. 2017

[42] King T, Cole M, Farber JM, Eisenbrand G, Zabaras D, Fox EM, et al. Food safety for food security: Relationship between global megatrends and developments in food safety. Trends in Food Science and Technology. 2017;**68**:160-175

[43] Jones FT. A review of practical Salmonella control measures in animal feed. Journal of Applied Poultry Research. 2011;**20**(1):102-113

[44] The European Commission. COMMISSION REGULATION (EU) No 200/2010 of 10 March 2010 implementing Regulation (EC) No 2160/2003 of the European Parliament and of the Council as regards a Union target for the reduction of the prevalence of Salmonella serotypes in adult breeding flocks. Official Journal of the European Union 2010;9

[45] The European Commission. COMMISSION REGULATION (EU) No 1190/2012 of 12 December 2012 concerning a Union target for the reduction of Salmonella Enteritidis and Salmonella Typhimurium in flocks of turkeys, as provided for in Regulation (EC) No 2160/2003 of the European Parliament a. Official Journal of the European Union 2012;29-34

[46] Anonymous. Commission regulation EU No 517/2011 implementing Regulation (EC) No 2160/2003 of the European Parliament and of the Council as regards a Union target for the reduction of the prevalence of certain Salmonella serotypes in laying hens of Gallus gallus and. Official Journal of the European Union 2011. p. L138/45.

[47] Correia-Gomes C, Leonard F, Graham D. Description of control programmes for Salmonella in pigs in Europe. Journal of Food Safety. 2021;**41**(5):e12916

[48] Ågren ECC, Lewerin SS, Frössling J. Evaluation of herd-level sampling strategies for control of Salmonella in Swedish cattle. Journal of Dairy Science. 2018;**101**(11):10177-10190

[49] Autio T, Tuunainen E, Nauholz H, Pirkkalainen H, London L, Pelkonen S. Overview of control programs for Non-eu-regulated Cattle Diseases in Finland. Frontier in Veterinary Science. 2021;**8**:778

[50] Santman-Berends IMGA, Mars MH, Weber MF, van Duijn L,

Waldeck HWF, Biesheuvel MM, et al. Control and eradication programs for non-EU Regulated Cattle Diseases in the Netherlands. Frontier in Veterinary Science. 2021;**8**:950

[51] Wegener HC, Hald T, Wong DLF, Madsen M, Korsgaard H, Bager F, et al. Salmonella Control Programs in Denmark. Emerging Infectious Diseases. 2003;**9**(7):774

[52] OIE - World Organisation for Animal Health. Prevention and Control of Salmonella in Commercial Pig Production Systems. In: Terrestrial Animal Health Code. France, Paris; 2018

[53] OIE. Biosecurity Procedures in Poultry Production. World Organisation for Animal Health. France, Paris; 2019. pp. 1-6

[54] Carrasco E, Morales-Rueda A, María G-GR. Cross-contamination and recontamination by Salmonella in foods: A review. FRIN. 2012;**45**:545-556

[55] Chang TFM, Iseppi L. EU Agro-Food chain and vertical integration potentiality: A strategy for diversification? Transition Studies Review. 2012;**19**(1):107-130

[56] Regmi A. Changing structure of global food consumption and trade. Market and Trade Economics Division, Economic Research Service. 2001;**2001**:1-3

[57] James HS, Klein PG, Sykuta ME. The adoption, diffusion, and evolution of organizational form: Insights from the agrifood sector. Managerial and Decision Economics. 2011;**32**(4):243-259

[58] Vestby LK, Møretrø T, Langsrud S, Heir E, Nesse LL. Biofilm forming abilities of Salmonella are correlated with persistence in fish meal- and feed factories. BMC Veterinary Research. 2009;**2009**:5

[59] Huss AR, Cochrane RA, Deliephan A, Stark CR, Jones CK. Evaluation of a biological pathogen decontamination protocol for animal feed mills. Journal of Food Protection. 2015;**78**(9):1682-1688

[60] Davies RH, Hinton MH. Salmonella in animal feed. In: Wray C, Wray A, editors. Salmonella in Domestic Animals. New York, NY: CAB International; 2000. pp. 285-300

[61] Muckey M, Huss AR, Jones C. Evaluation of liquid and dry chemical treatments to reduce salmonella typhimurium contamination on animal food manufacturing surfaces. Journal of Food Protection. 2022;**85**(5):792-797

[62] Møretrø T, Vestby LK, Nesse LL, Storheim SE, Kotlarz K, Langsrud S. Evaluation of efficacy of disinfectants against Salmonella from the feed industry. Journal of Applied Microbiology. 2009;**106**(3):1005-1012

[63] Wierup M, Widell S, Agr M. Estimation of costs for control of Salmonella in high-risk feed materials and compound feed. Infection Ecology & Epidemiology. 2014;**4**(1):23496

[64] Wedderkopp A, Gradel KO, Jorgensen JC, Madsen M. Pre-harvest surveillance of Campylobacter and Salmonella in Danish broiler flocks: A 2-year study. International Journal of Food Microbiology. 2001;**68**(1-2):53-59

[65] Davies R, Breslin M, Corry JE, Hudson W, Allen VM. Observations on the distribution and control of Salmonella species in two integrated broiler companies. The Veterinary Record. 2001;**149**(8):227-232

[66] Kloska F, Casteel M, Kump FWS, Klein G. Implementation of a risk-orientated hygiene analysis for the control of Salmonella JAVA in the broiler production. Current Microbiology. 2017;**74**(3):356-364

[67] Marin C, Balasch S, Vega S, Lainez M. Sources of Salmonella contamination during broiler production in Eastern Spain. Preventive Veterinary Medicine. 2011;**98**(1):39-45

[68] Castañeda-Gulla K, Sattlegger E, Mutukumira AN. Persistent contamination of salmonella, campylobacter, escherichia coli, and staphylococcus aureus at a broiler farm in New Zealand. Canadian Journal of Microbiology. 2020;**66**(3):171-185

[69] Luyckx KY, Van Weyenberg S, Dewulf J, Herman L, Zoons J, Vervaet E, et al. On-farm comparisons of different cleaning protocols in broiler houses. Poultry Science. 2015;**94**(8):1986-1993

[70] Davies R, Breslin M. Observations on Salmonella contamination of commercial laying farms before and after cleaning and disinfection. The Veterinary Record. 2003;**152**(10):283-287

[71] Wales A, Breslin M, Davies R. Assessment of cleaning and disinfection in Salmonella-contaminated poultry layer houses using qualitative and semi-quantitative culture techniques. Veterinary Microbiology. 2006;**116**(4):283-293

[72] Wales A, Breslin M, Carter B, Sayers R, Davies R. A longitudinal study of environmental salmonella contamination in caged and free-range layer flocks. Avian Pathology. 2007;**36**(3):187-197

[73] Carrique-Mas JJ, Marín C, Breslin M, McLaren I, Davies R. A comparison of the efficacy of cleaning and disinfection methods in eliminating Salmonella spp. from commercial egg laying houses. Avian Pathology. 2009;**38**(5):419-424

[74] Martelli F, Gosling RJ, Callaby R, Davies R. Observations on Salmonella contamination of commercial duck farms before and after cleaning and disinfection. Avian Pathology. 2017;**46**(2):131-137

[75] Marin C, Hernandiz A, Lainez M. Biofilm development capacity of Salmonella strains isolated in poultry risk factors and their resistance against disinfectants. Poultry Science. 2009;**88**(2):424-431

[76] Martelli F, Lambert M, Butt P, Cheney T, Tatone FA, Callaby R, et al. Evaluation of an enhanced cleaning and disinfection protocol in Salmonella contaminated pig holdings in the United Kingdom. PLoS One. 2017;**12**(6):e0178897

[77] De Busser EV, De Zutter L, Dewulf J, Houf K, Maes D. Salmonella control in live pigs and at slaughter. Veterinary Journal. 2013;**196**(1):20-27

[78] Wales AD, McLaren IM, Bedford S, Carrique-Mas JJ, Cook AJC, Davies RH. Longitudinal survey of the occurrence of Salmonella in pigs and the environment of nucleus breeder and multiplier pig herds in England. The Veterinary Record. 2009;**165**(22):648-657

[79] Lynch H, Walia K, Leonard FC, Lawlor PG, Manzanilla EG, Grant J, et al. Salmonella in breeding pigs: Shedding pattern, transmission of infection and the role of environmental contamination in Irish commercial farrow-to-finish herds. Zoonoses and Public Health. 2018;**65**(1):e196-e206

[80] Wirtanen G, Salo S. Cleaning and disinfection. In: Ninios T, Lundén J,

Korkeala H, Fredriksson-Ahomaa M, editors. Meat Inspection and Control in the Slaughterhouse. Chichester; 2014. pp. 453-472

[81] Zeng H, De Reu K, Gabriël S, Mattheus W, De Zutter L, Rasschaert G. Salmonella prevalence and persistence in industrialized poultry slaughterhouses. Poultry Science. 2021;**100**(4):100991

[82] De Busser EV, Maes D, Houf K, Dewulf J, Imberechts H, Bertrand S, et al. Detection and characterization of Salmonella in lairage, on pig carcasses and intestines in five slaughterhouses. International Journal of Food Microbiology. 2011;**145**(1):279-286

[83] Walia K, Argüello H, Lynch H, Grant J, Leonard FC, Lawlor PG, et al. The efficacy of different cleaning and disinfection procedures to reduce Salmonella and Enterobacteriaceae in the lairage environment of a pig abattoir. International Journal of Food Microbiology. 2017;**246**:64-71

[84] Arguello H, Álvarez-Ordoñez A, Carvajal A, Rubio P, Prieto M. Role of slaughtering in Salmonella spreading and control in pork production. Journal of Food Protection. 2013;**76**(5):899-911

[85] Rasschaert G, Houf K, Godard C, Wildemauwe C. Contamination of Carcasses with Salmonella during Poultry Slaughter. Journal of Food Protection. 2008;**71**(1):146-152

[86] Reiter MGR, Fiorese ML, Moretto G, López MC, Jordano R. Prevalence of Salmonella in a Poultry Slaughterhouse. Journal of Food Protection. 2007;**70**(7):1723-1725

[87] Buncic S, Sofos J. Interventions to control Salmonella contamination during poultry, cattle and pig slaughter. Food Research International. 2012;**45**(2):641-655

[88] Chmielewski RAN, Frank JF. Biofilm formation and control in food processing facilities. Comprehensive Reviews in Food Science and Food Safety. 2003;**2**(1):22-32

[89] Wales A, Taylor E, Davies R. Review of food grade disinfectants that are permitted for use in egg packing centres. World's Poultry Science Journal. 2021;**78**(1):231-260

[90] Corcoran M, Morris D, De Lappe N, O'Connor J, Lalor P, Dockery P, et al. Commonly used disinfectants fail to eradicate Salmonella enterica biofilms from food contact surface materials. Applied and Environmental Microbiology. 2014;**80**(4):1507-1514

[91] Giaouris E, Chorianopoulos N, Skandamis P, Nychas G-J. Attachment and biofilm formation by Salmonella in food processing environments. Salmonella – A Danger Foodborne Pathogens. 2012;**2012**:157-180

[92] Smith J, Corkran S, McKee SR, Bilgili SF, Singh M. Evaluation of post-chill applications of antimicrobials against Campylobacter jejuni on poultry carcasses. Journal of Applied Poultry Research. 2015;**24**(4):451-456

[93] Loretz M, Stephan R, Zweifel C. Antibacterial activity of decontamination treatments for pig carcasses. Food Control. 2011;**22**(8):1121-1125

[94] Wessels S, Ingmer H. Modes of action of three disinfectant active substances: A review. Regulatory Toxicology and Pharmacology. 2013;**67**(3):456-467

[95] Hahn S, Schneider K, Gartiser S, Heger W, Mangelsdorf I. Consumer exposure to biocides - identification

of relevant sources and evaluation of possible health effects. Environmental Health. 2010;**9**(1):7

[96] Anderson SE, Meade BJ. Potential health effects associated with dermal exposure to occupational chemicals. Environmental Health Insights. 2014;**8**(Suppl. 1):51

[97] Maillard J-Y. Antimicrobial biocides in the healthcare environment: Efficacy, usage, policies, and perceived problems. Therapeutics and Clinical Risk Management. 2005;**1**(4):307

[98] Thakur D, Ganguly R. Biocides. Environmental Micropollutants. 2022;**2022**:81-90

[99] Langsrud S, Sidhu MS, Heir E, Holck AL. Bacterial disinfectant resistance—a challenge for the food industry. International Biodeterioration & Biodegradation. 2003;**51**(4):283-290

[100] Tong C, Hu H, Chen G, Li Z, Li A, Zhang J. Disinfectant resistance in bacteria: Mechanisms, spread, and resolution strategies. Environmental Research. 2021;**1**:195

[101] Møretrø T, Heir E, Nesse LL, Vestby LK, Langsrud S. Control of Salmonella in food related environments by chemical disinfection. Food Research International. 2012;**45**(2):532-544

[102] Randall LP, Cooles SW, Coldham NG, Penuela EG, Mott AC, Woodward MJ, et al. Commonly used farm disinfectants can select for mutant Salmonella enterica serovar Typhimurium with decreased susceptibility to biocides and antibiotics without compromising virulence. The Journal of Antimicrobial Chemotherapy. 2007;**60**(6):1273-1280

[103] Karatzas KAG, Randall LP, Webber M, Piddock LJV, Humphrey TJ, Woodward MJ, et al. Phenotypic and proteomic characterization of multiply antibiotic-resistant variants of salmonella enterica serovar typhimurium selected following exposure to disinfectants. Applied and Environmental Microbiology. 2008;**74**(5):1508

[104] Webber MA, Randall LP, Cooles S, Woodward MJ, Piddock LJV. Triclosan resistance in Salmonella enterica serovar Typhimurium. The Journal of Antimicrobial Chemotherapy. 2008;**62**(1):83-91

[105] Wales AD, Davies RH. Co-selection of resistance to antibiotics, biocides and heavy metals, and its relevance to foodborne pathogens. Antibiotics. 2015;**4**(4):567

[106] Fernández Márquez ML, Burgos MJG, Pulido RP, Gálvez A, López RL. Biocide tolerance and antibiotic resistance in Salmonella isolates from Hen Eggshells. Foodborne Pathogens and Disease. 2017;**14**(2):89

[107] WebberMA,WhiteheadRN,MountM, Loman NJ, Pallen MJ, Piddock LJV. Parallel evolutionary pathways to antibiotic resistance selected by biocide exposure. The Journal of Antimicrobial Chemotherapy. 2015;**70**(8):2241

[108] Beier RC, Bischoff KM, Poole TL. Disinfectants (Biocides) used in animal production: Antimicrobial resistance considerations. In: Preharvest and Postharvest Food Safety: Contemporary Issues and Future Directions. Ames, Iowa, USA: John Wiley & Sons, Ltd; 2008. pp. 227-238

Chapter 7

Antimicrobial Resistance in *Salmonella*: Its Mechanisms in Comparison to Other Microbes, and The Reversal Effects of Traditional Chinese Medicine on Its Resistance

Hongxia Zhao

Abstract

Salmonella is one of the most notable pathogens leading to the outbreak of foodborne diseases worldwide. Antimicrobial chemotherapy with 3rd-generation cephalosporins or fluoroquinolones is often used for severe infections caused by *Salmonella*. Therefore, antibiotic or antimicrobial resistance (AMR) of *Salmonella* is a serious threat to human and animal health in China and worldwide. In order to better understand the current situation and development status of AMR in *Salmonella* isolates, this chapter will provide an overview of the following: 1. The history and development trend of AMR in *Salmonella*, and a comparison of its AMR with that of other major pathogenic bacteria in animals. 2. The AMR mechanisms of *Salmonella* to various antibiotics, with a particular focus on the commonly used antibiotics. 3. The mechanisms of the spread of AMR in *Salmonella*, including the AMR genes or mobile genetic elements carrying AMR genes among microbes, and among people, animal-derived foods, and the environment. 4. The elimination or reversal of AMR in *Salmonella* by using traditional Chinese medicine or the active ingredients in traditional Chinese medicine. 5. The development of detection technology for *Salmonella* serotypes, virulence, and AMR, and the improvement from conventional detection methods to more advanced biological detection methods and bioinformatics technology.

Keywords: antimicrobial resistance (AMR), *Salmonella*, salmonellosis in human and animals, comparison with other bacterial species, elimination and reversal of AMR, traditional Chinese medicine

1. Introduction

1.1 *Salmonella* and Salmonellosis

Salmonella is a spore-free, capsule-free, gram-negative straight bacilli, which widely exists in human and animal intestines. Genus *Salmonella* currently has two species, *Salmonella enterica* and *Salmonella bongori*. The type species, *S. enterica*, can be further classified into six subspecies with Roman numerals based on their genomic relatedness and biochemical properties, namely, I, *S. enterica* subsp. *enterica*; II, *S. enterica* subsp. *salamae*; IIIa, *S. enterica* subsp. *arizonae*; IIIb, *S. enterica* subsp. *diarizonae*; IV, *S. enterica* subsp. *houtenae*; and VI, *S. enterica* subsp. *indica* [1–3]. So far, *S. bongori* (V) has 22 serotypes [4]., and *S. enterica* has approximately 2600 different serotypes or serovars [2, 5]. *Salmonella enterica* subsp. *enterica* (I) is present predominantly in mammals and contributes approximately 99% of *Salmonella* infections in humans and warm-blooded animals. The other five *Salmonella enterica* subspecies and *S. bongori* are mainly found in environment and cold-blooded animals [4, 6]. Among human isolates, *S.* Enteritidis is the most common serotype, accounting for 65% of all isolates, and *Salmonella enterica* serovar Typhimurium was reported most frequently among nonhuman isolates, although no serotype predominated [7].

People usually get salmonellosis by eating contaminated foods, particularly foods of animal origin, or by direct contact with infected animals. *Salmonella* infection causes diarrhea, fever, vomiting and abdominal cramps. Salmonellosis is a common zoonotic disease. It could not only cause serious economic losses in animal production, but also a serious threat to human health [8]. Infection with *Salmonella enterica* usually results in diarrhea, fever, and abdominal cramps, but some people become asymptomatic or chronic carrier as a source of infection for others.

Salmonella is one of the most notable pathogens leading to the outbreak of foodborne diseases worldwide [8–10]. In the United States and other developed countries, the annual incidence rate of *Salmonella* infection is as high as 15.4% [11], and the disease outbreak and hospitalization caused by *Salmonella* are higher than those caused by other foodborne bacteria [12]. In China, about 300 million people are infected with *Salmonella* every year [13]. Salmonellosis accounts for 70% - 80% of the total number of foodborne diseases every year in China, and seriously threatens food safety and human health. In one report, 88 *Salmonella* strains were collected from patients and asymptomatic people in Nantong city of China from 2017 to 2018 [14]. Among these strains, 20 serotypes belonging to 8 serogroups were identified. *Salmonella typhimurium* remained to be the predominant serotype in strains from both patients and asymptomatic people. Among the 27 strains from patients, *S. enteritidis* and *S.* Rissen were shown as the other two major serotypes, while *S.* London, *S.* Derby, and *S.* Meleagridis were demonstrated as the other significant serotypes among the 61 strains from asymptomatic people. AMR testing revealed that 84.1% of strains from both resources were multi-drug resistant. By comparing the characteristics of *Salmonella* strains from two different kinds of sources, effective strategies would be developed to control *Salmonella* infection in humans.

Typhoid fever caused by typhoid bacilli is a human acute intestinal infection transmitted between humans. Fowl typhoid is mainly caused by *S. typhimurium*. Salmonellosis in cattle is mainly caused by *S. typhimurium* and *S. dublin*. It mainly occurs in calves aged 10–30 days, and dysentery is the main symptom, so it is also called calf paratyphoid. It is reported that *Salmonella spp.* are among the most important foodborne pathogens and the third leading cause of human death among diarrheal diseases worldwide [15]. Animals are the primary source of this pathogen, and animal-based foods are the main transmission route to humans. Thus, understanding the global epidemiology of *Salmonella* serovars is key to controlling and monitoring this bacterium. The study conducted by Rafaela *et al*. evaluated the prevalence and diversity of *Salmonella* serovars in animal-based foods (beef, pork, poultry, and seafood) throughout the five continents (Africa, the Americas, Asia, Europe, and Oceania) [15]. The results showed *S. typhimurium* presented a cosmopolitan distribution in all four assessed matrices and continents. Poultry continues to play a central role in the dissemination of *S. enteritidis* serovar to humans, and *S*. Anatum and *S*. Weltevreden were the most frequently found in beef and seafood, respectively. Careful monitoring of certain serovars and the main vehicles for the transmission of this pathogen will promote the improvement of control programs to reduce the risk of this pathogen reaching humans.

The dominant serotypes of *Salmonella* from different countries and animals are different. The serotypes of *Salmonella* from American chickens are mainly from Kentucky [16]. The predominant serotype of *Salmonella* from cattle in Iran is *S. typhimurium* [17]. The serotypes of *Salmonella* from chickens in China are mainly *S*. Enteriditis, *S*. Pullorum, and *S. typhimurium* [18]. In terms of the serotyping of *Salmonella*, the conventional detection method is to determine the O antigen and H antigen by slide agglutination, and then determine the serotype according to the serum antigen table. Antibiotics have been used in clinical treatment for more than half a century. Antimicrobial therapy of infections based on the antibiotic susceptibility test results and type plays an important role in prevention and treatment of Salmonellosis.

1.2 Antimicrobial-resistance of *Salmonella* and the effect of traditional Chinese medicines on antibiotic-resistant *Salmonella*

The overall antibiotic or antimicrobial resistance (AMR) of *Salmonella* increased significantly from 20% ~ 30% in the early 1990s to 70% at the beginning of this century [8]. Different serotypes show different AMR to antibiotics, and the AMR rate to different antibiotics is also different [9–12]. In the past three decades, the drug resistance of *Salmonella* has been significantly enhanced, accompanied by the continuously widened spectrum of multiple AMRs. At present, the antibiotics used for *Salmonella* are mainly β-lactams, aminoglycosides, sulfonamides, macrolides, phenylpropanols, quinolones, and tetracyclines [19]. With the increasing dosage and abuse of antibiotics, the AMR of *Salmonella* is becoming more and more prominent. The irrational use of antibiotics has led to a gradual increase in AMR of animal-derived pathogens. From the overall situation of China, China has become one of the countries with the most serious AMR of animal-derived bacteria in the world. The AMR is becoming more and more serious and leads to the effect of clinical treatment decreasing or failing. Multidrug-resistant strains are regionally prevalent and can be transmitted along the food chain, posing risks to food safety and human health.

Different serotypes show different AMR to antibiotics, and the drug resistance rate to different antibiotics is also different [9–12]. In recent years, *Salmonella* which has shown resistance to quinolones (ciprofloxacin) and the third-generation cephalosporins (ceftriaxone, cefotaxime) has been reported in China, France, and other countries and regions [20–23], indicating that with the wide clinical application, the therapeutic effect of ideal antibiotics is also declining. The AMR can be encoded by endogenous AMR genes, or generated by gene mutation or acquisition of exogenous

AMR genes carried by mobile genetic elements. Among them, the exogenous AMR genes carried by plasmids, Integron (In), bacteriophages, and Transposon (Tn) can be horizontally transferred through transformation, transduction, and conjugation, which is the main reason for the rapid spread of acquired AMR of bacteria [24].

Different serotypes of *Salmonella* have different AMR [25], and the rise of AMR levels also brings severe challenges to the prevention and treatment of salmonellosis [26]. Therefore, accurate and rapid serotype identification and AMR detection are of great significance for the prevention and control of salmonellosis [27, 28]. Therefore, how to quickly and efficiently identify the serotype and AMR of *Salmonella* has become an urgent practical problem, and the introduction of new detection methods is imperative.

Some traditional Chinese medicines have the following properties: anti-bacterial, anti-inflammatory, nourishing and improving immunity, low potential for building tolerance, and low toxicity and side effects. Some studies have shown that traditional Chinese medicine can eliminate AMR plasmids, have a reversal effect on bacterial resistance, and reduce the selection pressure of bacteria [29, 30]. Therefore, as an alternative to antimicrobial agents or a promoter of antimicrobial agents, it has become one of the research hotspots, which has important significance for the prevention and treatment of *Salmonella* infectious diseases.

1.3 The objective of the chapter

The AMR of *Salmonella* isolates from humans and animals is becoming more and more serious, which creates great difficulties in the prevention and control of infectious diseases caused by resistant *Salmonella* isolates. *Salmonella* with AMR can not only spread widely between animals but also through food to infect humans. Moreover, AMR can also be passed to humans, So it is a huge potential threat to human and animal health in China and worldwide. To make people pay more and more attention to the problem of AMR in *Salmonella,* this chapter will first review the history and developing trend of AMR in *Salmonella*. The occurrence and spread mechanisms of the AMR of *Salmonella* will be clarified, to provide a theoretical basis for searching a new efficacious antibiotic to eliminate and weaken its resistance and control of Salmonellosis caused by resistant *Salmonella.* In addition, the whole genome sequencing technology has high accuracy in predicting the serotype and AMR of *Salmonella*. The advanced biological detection methods and bioinformatics technology used in identifying *Salmonella* serotypes and AMR will be introduced in this chapter. They have broad application prospects in determining *salmonella* serotype and AMR and the results for prediction will play a very important part in providing strong guidance for the rational use of antibiotics in the clinic.

2. The history and developmental trend of AMR in *Salmonella*, and a comparison of its AMR with that of other major animal-derived pathogenic bacteria

2.1 Development trend of AMR in *Salmonella*

Salmonella is one of the most common agents of gastrointestinal disease globally. In the United States, nontyphoidal *Salmonella* is the second most frequent bacterium causing foodborne illness and the first bacterial pathogen in terms of hospitalizations

DOI: http://dx.doi.org/10.5772/intechopen.113376

and deaths. For severe infections, antimicrobial chemotherapy with 3rd-generation cephalosporins or fluoroquinolones is recommended. Therefore, AMR in *Salmonella* is considered a serious public health threat. 22,102 genomes from public databases were analyzed to track AMR trends in nontyphoidal *Salmonella* in food animals in the United States. In 2018, genomes deposited in public databases carried genes conferring resistance, on average, to 2.08 antimicrobial classes in poultry, 1.74 in bovines, and 1.28 in swine. There was a decline in AMR of over 70% compared to the levels in 2000 in bovines and swine and an increase of 13% for poultry. Trends in resistance inferred from genomic data showed good agreement with U.S. phenotypic surveillance data. In 2018, resistance to 3rd-generation cephalosporins in bovines, swine, and poultry decreased to 9.97% on average, whereas in quinolones and 4th-generation cephalosporins, resistance increased to 12.53% and 3.87%, respectively.

At present, the antibiotics used for *Salmonella* are mainly β-lactams, aminoglycosides, sulfonamides, macrolides, phenylpropanols, quinolones, and tetracyclines. The β-lactam mainly includes penicillins (such as ampicillin, carbenicillin, etc.), β-lactam enzyme inhibitors (such as amoxicillin-clavulanic acid, ampicillin-sulbactam, etc.), and cephalosporins (such as ceftriaxone, cefoxitin, etc.). Other antibiotics mainly include aminoglycosides (such as gentamicin, kanamycin, etc.), sulfonamides (such as sulfamethoxazole, trimethoprim-sulfamethoxazole, etc.), macrolides (such as azithromycin, etc.), phenylpropanols (such as chloramphenicol, etc.), quinolones (such as nalidixic acid, ciprofloxacin, etc.) and tetracyclines (such as doxycycline, tetracycline) [18]. With the increasing dosage and abuse of antibiotics, the drug resistance of *Salmonella* is becoming more and more prominent. Such as the prevention and treatment of the decline, the emergence of new drug resistance genes, and multidrug resistance (MDR). *Salmonella* as a zoonosis, the enhancement of AMR is also seriously endangering human health and safety [21]. *Salmonella* resistance to a single antibiotic first appeared in the 1960s [22]. Subsequently, AMR of *Salmonella* emerged in different countries and regions of the world, and the isolation rate increased accordingly. In research by Khan *et al.* [23], the isolation rate of MDR of *Salmonella typhi* was higher in Asia and Africa. The results showed that the isolation rates in India, Pakistan, and Vietnam were significantly higher than those in Indonesia and China. Fluoroquinolones and cephalosporins are currently the preferred antibiotics for clinical prevention and control of *Salmonella* infection, but with the irregular use of fluoroquinolones and cephalosporins, the AMR spectrum of *Salmonella* is wider, and there is a large degree of cross-resistance. Hasan *et al.* [31] showed among MDR *Salmonella*, *S. paratyphi* showed a higher level of resistance to fluoroquinolones. *Salmonella* strains isolated from animal-derived foods have a high level of resistance to tetracycline. Generally, the resistance rate can reach 80%, and can even reach a high level of 85%. It shows a certain level of resistance to chloramphenicol, penicillin, nalidixic acid, and sulfonamide antibiotics. In addition, the problem of multi-drug resistance is also very serious. The resistance rate to two or more antibiotics can reach 75%, and the resistance rate to five or more antibiotics can reach 30% [24, 32, 33]. The resistance level of *Salmonella* differs between different studies and regions. Clinically isolated *Salmonella* strains showed a high level of resistance to nalidixic acid, ampicillin, chloramphenicol, and other antibiotics (65% -90%), and the resistance level to sulfonamides, tetracycline, streptomycin was around 50%, and the resistance to the second and third generation cephalosporins was lower, can reach 10% [34, 35].

At present, the problem of AMR of pathogenic bacteria in veterinary clinics is becoming more and more serious [36]. To promote the growth of livestock and poultry, there will be a large number of antimicrobials used, and many veterinary surgeons in the clinical treatment of antibiotics for the irrational use of non-standard, resulting in a gradual increase in the level of *Salmonella* resistance, multi-drug resistance is becoming increasingly serious [37]. Changes in the resistance spectrum occur as *Salmonella* mutates in the natural environment and clinical treatments and are the result of bacterial evolution [38]. *Salmonella* isolates from clinical specimens have been increasing in recent years, and AMR rates are rising rapidly around the world [39]. With the introduction of new antibiotics into clinical use, the corresponding AMR strains will also be rapidly produced, and single AMR has gradually developed into multidrug resistance. The problem of AMR has become more and more serious, and the problem of bacterial resistance has been paid more and more attention [40, 41]. *Salmonella* resistance can not only spread widely between animals but also through food to infect humans, causing food poisoning. AMR can also be passed to humans, affecting human health [42]. The increasingly serious AMR of *Salmonella* has had a great impact on the efficacy of traditional antibiotics, and the increase in the resistance of *Salmonella* strains to new antibiotics has had a more adverse effect on clinical treatment.

2.1.1 Resistance to tetracyclines

Tetracycline antibiotics are broad-spectrum antibiotics produced by actinomycetes and contain a fused tetraphenyl ring structure [43]. They can be used to treat bacterial diseases caused by Gram-positive and Gram-negative bacteria. Tetracycline antibiotics are mainly divided into two categories: natural and semi-synthetic antibiotics, mainly chlortetracycline, oxytetracycline, methacycline, doxycycline, dimethylaminotetracycline, etc. Due to the characteristics of tetracycline antibiotics, livestock and poultry can only absorb part of them. Most antibiotics will enter the breeding environment in the form of antibiotics themselves or metabolites through the way of livestock and poultry excreta. In addition, livestock and poultry are closely related to human beings. With the continuous development of animal husbandry, bacterial diseases have become increasingly prominent in both intensive farming and free-range farming, and prevention and treatment are facing tremendous pressure. In the prevention or treatment of bacterial diseases, antibiotics are often used. However, when antibiotics are used, there is excessive use, misuse, and abuse, which leads to the specific selection of pathogenic microorganisms by antibiotics and the resistance of pathogenic microorganisms. Among these pathogenic microorganisms, *Salmonella* is more resistant to tetracycline antibiotics. The resistance of *Salmonella* to tetracycline antibiotics varies from country to country, which is related to the unreasonable use of tetracycline antibiotics.

Zhang [44] isolated and identified 34 strains of *Salmonella* from three breeding chicken farms in eastern Liaoning Province. After an antibiotic sensitivity test, 30 of them were resistant to tetracycline. Di *et al.* [45] found that the resistance rate of swine *Salmonella* to oxytetracycline was as high as 58.3%. Li *et al.* [46] found that the resistance rates to doxycycline and oxytetracycline in 247 strains of *Salmonella* isolated from pigs were as high as 89.77% and 94.88%, respectively. The strains showing resistance to doxycycline and oxytetracycline were as high as 89.3%. This shows that *Salmonella* is not only resistant to single tetracycline antibiotics but also resistant to two or more tetracycline antibiotics. The continuous emergence of high resistance rates indicates that tetracycline antibiotics are used too much and too frequently in the clinical treatment of avian salmonellosis. The use of tetracycline antibiotics should be appropriately reduced or replaced.

2.1.2 Resistance to quinolone

Quinolone antibiotics, also known as pyruvic acid or pyridine copper acid antibiotics, are a class of synthetic antibiotics with 4-quinolone, which mainly inhibit gram-negative bacteria and mycoplasma. Quinolones have been used to treat human and animal infectious diseases and promote animal growth because of their broad antimicrobial spectrum, strong bactericidal effect, rapid action, lack of cross-resistance with other antibiotics, and few side effects [47]. The common quinolones in clinical treatment are enrofloxacin, ciprofloxacin, ofloxacin, sarafloxacin, difloxacin, and so on.

Yao *et al.* [48] found that the resistance rate of *Salmonella* isolated from Shanxi Province, China to the first-generation quinolones was the highest, reaching 56.93%. Zhang *et al.* [49] found that 2.33% (34 of 1523) of *Salmonella enteritidis* strains were resistant to ciprofloxacin. Among them, 11 strains had high resistance to ceftriaxone, and all ciprofloxacin-positive strains had resistance to at least 7 antibiotics. From 2013 to 2018, the resistance rate of *Salmonella enteritidis* to fluoroquinolone enrofloxacin (8.50% -16.30%) showed an increasing trend year by year. In 2012, Li *et al.* [50] conducted an antibiotic sensitivity test on 62 strains of *Salmonella* isolated from pigs. The results showed that the resistance rate of fluoroquinolones was 88.7%. As one of the main antibiotics for the treatment of *Salmonella*, quinolones still have an increasing resistance rate year by year, which has become the hardest hit area of *Salmonella* resistance.

2.1.3 Resistance to aminoglycosides

There are many kinds of aminoglycoside antibiotics. The earliest aminoglycoside antibiotic is streptomycin, followed by gentamicin, kanamycin, spectinomycin, neomycin, amikacin, netilmicin, and so on. Aminoglycoside antibiotics are mainly divided into two categories: natural and semi-synthetic. Natural aminoglycoside antibiotics include streptomycin, kanamycin, tobramycin, neomycin, spectinomycin, gentamicin, etc. Semi-synthetic aminoglycoside antibiotics include amikacin, netilmicin, etc. [51].

Because of their low price and remarkable effect, aminoglycoside antibiotics are widely used in the treatment and prevention of animal diseases in animal husbandry and aquaculture [52]. However, the use of aminoglycoside antibiotics is abused and abused, resulting in excessive antibiotic residues in animal bodies and AMR. Therefore, the use of aminoglycoside antibiotics has been limited by many countries [53]. Guan *et al.* [54] conducted an AMR test on 23 isolated and identified *Salmonella* strains. The results showed that the resistance rate to gentamicin was the highest, which was 66.7%. The resistance rate to spectinomycin was 33.3%, and the resistance rate to kanamycin and tobramycin was 16.7%. The 13 strains of *Salmonella* isolated by Zhang *et al.* [55] were tested for AMR to 10 commonly used antibiotics, all of which showed high AMR rates with resistance to more than two antibiotics. Some even achieved resistance to 8 of them, although sensitive to gentamicin and kanamycin. The AMR results varied among the 13 *Salmonella* isolates, possibly due to the changing breeding environment or AMR. Thus, in recent years, *Salmonella* resistance to aminoglycoside antibiotics has been very serious, and mostly multi-drug resistance.

2.1.4 Resistance to amide alcohols

Amide alcohol antibiotics are also called chloramphenicol antibiotics. They are a class of antibiotics with broad-spectrum antibacterial amide alcohol substances, which have inhibitory effects on both Gram-positive and negative bacteria. In the field of agriculture in animal husbandry, aquaculture, and chemical industry in the cosmetics industry are widely used, mainly for the treatment of chicken, pig, cattle, and other animals respiratory disease infections. Amide alcohol antibiotics mainly include chloramphenicol, palm chloramphenicol, succinomycin, florfenicol, thiamphenicol, etc.

In 2019, China explicitly banned the continued use of chloramphenicol in foodborne animals. At present, thiamphenicol and florfenicol are widely used as substitutes for chloramphenicol in animal husbandry. With the wide application of amide alcohol antibiotics, the resistance of *Salmonella* to amide alcohol antibiotics has gradually increased. Huang *et al.* [56] conducted an antibiotic resistance or AMR test on 61 isolated *Salmonella* strains. The results showed that the resistance rate to florfenicol accounted for 19.67%. Mondal *et al.* [57] conducted an AMR test on 9 isolated *Salmonella* strains. The results showed that 9 *Salmonella* strains were highly sensitive to ciprofloxacin, kanamycin, nalidixic acid, cotrimoxazole, and ampicillin, but highly resistant to chloramphenicol. Li *et al.* [58] conducted an AMR test on 215 strains of *Salmonella* isolated in Henan Province in China. The results showed that the resistance rate to florfenicol was 92.56%, and the AMR was serious. With the extensive use of florfenicol, the number of strains resistant to florfenicol showed an increase. Since February 2022, *Salmonella* strains resistant to florfenicol mainly belong to *S. typhimurium*, *S.* Agona, and *S. paratyphi*.

2.2 Comparison of AMR in *Salmonella* with other major animal-derived pathogens

China has become the world's largest producer and consumer of livestock and poultry products [58]. The production of pork, poultry meat, and eggs has been the world's first for several consecutive years, and milk production is the third in the world. The rapid growth of China's aquaculture industry mainly depends on the expansion of the scale of aquaculture and the increase in the number of aquaculture facilities. The large-scale and intensive aquaculture industry continues to develop steadily. Veterinary antibiotics, especially antibiotics, play an important role. However, the irrational use of antibiotics has led to a gradual increase in AMR of animal-derived pathogens. The sensitivity of animal-derived pathogens to quinolones, β-lactams, and other important antibiotics is declining, and the AMR is getting higher and higher. Some clinical isolates of pathogens are resistant to more than 15–20 kinds of antimicrobial agents, leading to livestock and poultry disease prevention and control becoming increasingly close to the embarrassing situation of no antibiotic being available [59]. *Streptococcus*, *Haemophilus parasuis*, *Pasteurella multocida* and other important animal-borne pathogens of amoxicillin, enrofloxacin, and other antimicrobial resistance are becoming more and more serious with clinical treatment, losing effectiveness or failing. In the breeding industry, for a long time, widely through mixing, drinking water to livestock and poultry use of antimicrobial, healthy animal intestinal symbiotic *Escherichia coli*, *Enterococcus* resistance to commonly used antibiotics is also increasing year by year [59]. The AMR of *Salmonella* from livestock and poultry is developing continuously, and the antimicrobial resistance mechanism is becoming more and more complex [60]. Multidrug-resistant strains are regionally prevalent and can be transmitted along the food chain, posing risks to food safety and human health. The emergence and prevalence of five AMR *cfr* genes have brought great challenges to the clinical treatment of methicillin-resistant *Staphylococcus aureus* (MRSA), and vancomycin-resistant *Enterococcus* infection [58]. The detection rate of

S. aureus clinical strains *cfr* in developed countries is less than 0.5%. The detection rate of *S. aureus* clinical strains *cfr* in China is much higher than that in developed countries by nearly 4%. This gene has even been found in animal-derived *Bacillus*, *Streptococcus*, *Enterococcus*, *Escherichia coli*, and *Proteus*, and is mostly located in plasmid DNA that can be horizontally transmitted [58].

Zhao *et al*. [59] isolated 4 main pathogenic bacteria from 260 cow endometritis samples in Inner Mongolia, including 126 strains of *E. coli* (48.5%), 84 strains of *Streptococcus* (32.3%), 53 strains of *S. aureus* (20.4%) and 21 strains of *Salmonella* (8.1%). The results of an antimicrobial susceptibility test showed that the resistance rate of *E. coli* to sulfonamides and benzylaminopyrimidines was more than 98%, and the resistance rate to ceftiofur was 13.7%. The resistance rate of *Streptococcus* to β-lactams, tetracycline, and kanamycin was more than 80%, and the resistance rate to vancomycin was 26.7%. The resistance rate of *S. aureus* to β-lactams ranged from 60–85%, to gentamicin and three combinations ranged from 7.5% to 1.2%, and was completely sensitive to vancomycin. The resistance rates of *Salmonella* to β-lactams, gentamicin, tetracyclines such as oxytetracycline and doxycycline were between 75% and 90%. *Salmonella* was sensitive to cefotaxime, and the resistance rate was 29%. The resistance rates to aminoglycosides such as tobramycin and amikacin were less than 10%. Four isolates were sensitive to fluoroquinolones and the resistance rates were less than 35%. Zhao *et al*. [60] isolated pathogenic bacteria from 40 samples of cow endometritis in Xinjiang mainly include *E. coli*, *Staphylococcus*, *Streptococcus*, *Bacillus cereus* and *Salmonella*, and the first three pathogens are the main pathogenic bacteria. The results of antibiotic sensitivity test showed that cefotaxime and amoxicillin had obvious antibacterial effect on *E. coli*, enrofloxacin and kanamycin had obvious antibacterial effect on *Staphylococcus*, and amoxicillin and ciprofloxacin had obvious antibacterial effects on *Streptococcus*. Almost all isolated bacteria were resistant to tetracycline and penicillin and were sensitive to quinolones and lactams. Based on the above studies, it seems that *Salmonella* showed different patterns of AMR to some commonly used antibiotics when compared with several other major animal-derived pathogenic bacterial species. The underlying mechanisms are not clear. They could be due to different bacterial niches, different standards for antibiotic usage and animal breeds. Further research is needed to explore the mechanisms which could be important for designing strategies for migrations of AMR.

3. The AMR mechanisms of *Salmonella* to various antibiotics, with a particular focus on the commonly used antibiotics

The extensive use of antibiotics has inevitably improved the survival adaptability of pathogenic bacteria and the endogenous flora of humans and animals, and promoted the evolution of their genomes, thus leading to the emergence and spread of AMR strains. At the beginning of this century, the overall AMR of *Salmonella* increased significantly from 20% ~ 30% in the early 1990s to 70% [29]. Different serotypes show different AMR to antibiotics, and the AMR rate also varies between different antibiotics [30, 61–63]. In recent years, *Salmonella,* which has shown resistance to quinolones (ciprofloxacin) and the third generation of cephalosporins (ceftriaxone, cefotaxime) has been reported in China, France, and other countries and regions [64–67], indicating that with the wide clinical application, the therapeutic effect of ideal antibiotics is also declining. In addition, the emergence and global spread of multi-antibiotic resistant *Salmonella* make the situation of AMR of *Salmonella* extremely severe. Therefore, the use of antibiotics should be further standardized and the AMR monitoring of *Salmonella* should be strengthened in the future.

The biochemical mechanisms of AMR can generally be classified into three categories [68–70]: 1) Produce inactivating enzymes to destroy antibacterial antibiotics through hydrolysis or modification, so that they can be converted into derivatives without antibacterial activity; 2) Reduce the permeability of the bacterial outer membrane, hinder the entry of antibacterial agents, or strengthen the efflux of active efflux pump to transport antibacterial agents out of the cell to reduce the antibiotic concentration in the cell; 3) To modify the action target of antibiotics or cause target mutation through gene mutation, thereby reducing the affinity of antibiotics to target proteins. The AMR can be encoded by endogenous AMR genes, or generated by gene mutation or acquisition of exogenous AMR genes carried by mobile genetic elements. Among them, the exogenous AMR genes carried by plasmids, Integron (In), bacteriophages, and Transposon (Tn) can be horizontally transferred through transformation, transduction, and conjugation, which is the major reason for the acquired AMR and rapid spread of bacteria [71].

Plasmids are extrachromosomal DNA molecules that can replicate autonomously and can confer host resistance to important antibiotics, including β-Lactamides, aminoglycosaminoamines, tetracyclines, chloramphenicols, sulfonamides, trimethoprims, macrolides and quinolones [72], and conjugated plasmids can transfer AMR to recipient bacteria through conjugation. Plasmids are closely related to the current situation of *Salmonella* resistance, and heavy metal resistance genes, disinfectant resistance genes, and virulence-related genes carried on plasmids have improved the survival adaptability of *Salmonella* to the environment [73].

Salmonella has a high level of resistance to quinolones, mainly due to the mutation of *gyrA*, *gyrB*, *parC* and *parE* genes in the quinolone resistance determining region (QRDR) on the bacterial chromosome, which makes the antibiotics lose their binding sites and efficacy [71]. The quinolone resistance genes *qnr*, *aac* (6′) - *Ib cr*, *qepA*, and *oqxAB* carried by plasmids can mediate low levels of quinolone resistance and accelerate the mutation of *gyrA*, *gyrB*, *parC*, and *parE* genes in QRDR, which is the main reason for the spread of quinolone resistance in *Salmonella* at present [67, 74].

The tolerance of *Salmonella* to β-lactam drugs is mainly due to the hydrolysis of antibacterial drugs β-lactamases, and most β-lactamase gene is carried by plasmid. Among them, plasmid-mediated ultra-broad spectrum β-lactamase genes *blaCTX-M*, *blaTEM*, and *blaSHV*, *AmpC* β-lactamase gene *blaCMY* and carbapenemase genes *blaKPC*, *blaVIM*, *blaIMP*, and *blaOXA* are prevalent worldwide [63, 71, 75].

In addition, the plasmid can also achieve the aggregation and transfer of antibiotic-resistant gene clusters by capturing mobile elements such as integrons or transposons. Integron is a natural cloning and expression system found in bacteria in recent years. Although the integron lacks the ability of autonomous movement, it often participates in the transfer as a component of the conjugated plasmid or transposon, thus promoting the diffusion of antibiotic-resistant genes [76]. Vo [77] detected *aadA1*, *aadA2*, *aadA5*, *blaPSE-1*, *blaOXA-30*, *dfrA1*, *dfrA12*, *dfrA17* and sat resistance gene cassettes in the type I integron carried by *Salmonella* isolates in Vietnam, forming nine different gene box arrays, and most of them are located on conjugative granules, which can transfer resistance to *E. coli* or *S. enteritidis* receptor bacteria.

4. The elimination or reversal of AMR in *Salmonella* by using traditional Chinese medicine or the active ingredients in traditional Chinese medicine

Chinese herbal medicine is natural and has many advantages: low toxicity, and lower residual levels of toxic substances [78]. It plays an active role in modern infection prevention and control. Some traditional Chinese medicines have the following properties: anti-bacterial, anti-inflammatory, nourishing and improving immunity, low potential for building tolerance, and low toxicity and side effects. Some studies have shown that traditional Chinese medicine can eliminate AMR plasmids, have a reversal effect on bacterial resistance, and reduce the selection pressure of bacteria [78, 79]. Therefore, as an alternative to antimicrobial agents or a promoter of antimicrobial agents, it has become a research hotspot, which has important significance for the prevention and treatment of *Salmonella* infectious diseases.

Some studies have shown that Chinese herbal medicines have a bacteriostatic effect on *Salmonella* in calves, and the bacteriostatic intensity ranked from strongest to weakest as follows: gallnut, schisandra chinensis, wumei, chebula, *Ligustrum lucidum*, pomegranate peel, sumu, and scutellaria. Among them, gallnut has the best bacteriostatic effect [79]. Wumei, coptis chinensis, and rhubarb have bacteriostatic effects on the intestinal *Salmonella* of dairy cows [80], among which, gallnut has good bacteriostatic effects on *S. typhimurium* and *S. cholera-suis* isolates from pigs [81]. Ma [82] found that the elimination rates of resistance to amoxicillin and tetracycline were 1% and 5%, respectively, in the resistant *Salmonella* treated with ebony. Cao [83] found the elimination effects of Galla Chinesis and Scutellaria on *Salmonella* AMR and with the highest removal rate of resistant strains 23.3%, 15.3% respectively by 20 hours, and 14.7%, 9.9% respectively by 48 hours. The Galla Chinesis and Scutellaria showed resistant plasmid removal rate of 15.6% and 10.8%, respectively.

5. The development of detection technology for *Salmonella* serotypes, virulence, and AMR, and the change from conventional detection methods to more advanced biological detection methods and bioinformatics technology

Different serotypes of *Salmonella* have different antimicrobial resistance [25], and the rise of AMR level also brings severe challenges to the prevention and treatment of salmonellosis [26]. Therefore, accurate and rapid serotype identification and AMR detection are of great significance for the prevention and control of salmonellosis [27, 28]. In terms of the serotyping of *Salmonella*, the conventional detection method is to determine the O antigen and H antigen by slide agglutination, and then determine the serotype according to the serum antigen table. This serotyping technique requires high serum quality, costs a lot, and takes a long time, and some agglutinations are not obvious and difficult to distinguish. In terms of AMR detection of *Salmonella*, the most commonly used method is the antibiotic sensitivity test recommended by the American Committee for Clinical Laboratory Standardization (CLSI) [84]. However, the accuracy of the experimental results of this method is easily affected by experimental materials, experimental conditions, and personnel operations. In terms of AMR gene detection, common PCR detection techniques cannot identify all AMR genes at once [85]. Therefore, how to quickly and efficiently identify the serotype and AMR of *Salmonella* has become an urgent practical problem, and the introduction of new detection methods is imperative.

With the increasing maturity of sequencing technology, rapid, low-cost, and cost-effective whole genome sequencing technology (WGS) has been widely used in the research of bacterial epidemiology [86]. At the same time, the development of bioinformatics technology has also promoted the creation of a variety of public databases such as the serological typing of foodborne pathogens and antibiotic-resistant genes, such as the SeqSero serotype database and ResFinder AMR gene database. With the continuous updating and improvement of the databases, the accuracy of automatic data analysis will be higher and higher. Several studies have shown that WGS has broad application prospects in determining *Salmonella* serotype and AMR genotype, and may replace conventional laboratory methods in the future [87, 88]. At present, there is very limited research in this field in China.

The establishment of serotype databases promotes the application of WGS in *Salmonella* serotyping. The commonly used serotype databases include SeqSero and SISIR. At present, SeqSero has been updated to SeqSero2, which improves the accuracy of the serotype database. Compared with the SISTR database, SeqSero2 does not need the help of genome-wide multi-site sequence typing research, simplifying the operation process and making the application more convenient [89]. Xu *et al.* [90] selected 38 *Salmonella* strains from the American *Salmonella* surveillance system, and the coincidence rate between the WGS typing results and the original results was 100%. Zhang *et al.* [91] conducted molecular analysis on 308 known *Salmonella* serotypes through WGS, among which 304 strains were completely consistent in serotype, with a coincidence rate of 98.7%. Diep *et al.* [92] collected 100 *Salmonella* strains from the Netherlands, and the serotypes of 98 *Salmonella* strains predicted by WGS were consistent with the conventional typing results, with a coincidence rate of 98.0%. Robertson *et al.* [93] extracted *Salmonella* WGS data from the SPA public database, and the coincidence rate between the identified serotype and the original results was 95.0%.

In conclusion, WGS typing method has high accuracy in predicting common serotypes. Compared with the conventional serum typing method, WGS typing is faster. For rare serotypes that require different culture media and antisera to determine flagella (H1 and H2), WGS takes only a few minutes, while the conventional serum typing method may take several weeks, sometimes requiring multiple repetitions. Therefore, the typing method based on WGS opens a new door for the identification of *Salmonella* serotypes, which has great application value in *Salmonella* serotyping. With the improvement of sequencing technology and the improvement of the databases, WGS typing is expected to become a new standard for *Salmonella* serotyping [94, 95].

The emergence of AMR is closely related to the existence of AMR genes, and the expression of AMR genes determines bacterial AMR. Research shows that the ResFinder resistance gene database can detect more resistance genes in the prediction of resistance genes, and it is the preferred tool for AMR analysis [96]. Neuert *et al.* [97] compared the AMR of 3415 *Salmonella* strains to 15 kinds of antibacterial agents, and their genotypes, and found 97.8% correlation.

Zankari *et al.* [98] predicted the AMR of 49 strains of *S. typhimurium* to 17 kinds of antibiotics, which was completely consistent with the results of AMR phenotype. Among 189 *Salmonella* strains studied by Zhu *et al.* [99], the coincidence rates of WGS AMR prediction to sulfamethoxazole, ampicillin, and tetracycline with their AMR phenotypes were 97.8%, 94.6% and 85.7%, respectively.

For antibiotics whose AMR genotype is not clear or is still under study, the coincidence rate between the AMR phenotype predicted by WGS and the AMR genotype is relatively low. The resistance mechanism of enrofloxacin and ceftiofur is mainly related to chromosome-mediated mutations. At present, WGS has only detected plasmid-mediated resistance genes, while the resistance genes generated by chromosome mutations have not been detected. This may be due to the low coverage of some regions in the genome sequencing process, preventing the detection of mutation sites, or the emergence of new resistance gene mutations [100].

Overall, the genome-based genotyping method avoids the influence of subjective judgment of conventional serotyping methods and has a high application prospect in serotyping. It is expected to replace conventional serotyping methods. The prediction of AMR by antibiotic resistant genotypes also provides a new perspective and method for clarifying AMR mechanisms and detecting AMR [101]. When new serotypes or AMR genes appear, they can be directly retrieved and analyzed through WGS data, without the need for routine bacterial culture and identification again, which provides a simpler method for the analysis of *Salmonella* serotypes and AMR. In addition, the application of WGS has also promoted research and development in other directions such as the genetic and variation characteristics of foodborne pathogens, AMR mechanisms [102], and will have an increasing impact on the analysis and research of the molecular biological characteristics of bacteria in different ecosystems and the substitution of traditional methods [103, 104]. With the development of whole gene sequencing technology and the reduction of its cost, rapid screening of antibiotic-resistant genes from genome data by bioinformatics methods has become a research hotspot.

6. Conclusion

The resistance of *Salmonella* to β-lactams, gentamicin, tetracyclines such as oxytetracycline and doxycycline was serious. However, *Salmonella* isolates were sensitive to fluoroquinolones, cefotaxime, and aminoglycosides such as tobramycin and amikacin. *Salmonella* has shown resistance to quinolones (ciprofloxacin) and the third generation cephalosporins (ceftriaxone, cefotaxime) in China, France, and other countries and regions. The resistance of *Salmonella* from livestock and poultry is developing continuously, and the AMR mechanism is becoming more and more complex. Multidrug-resistant *Salmonella* is regionally prevalent and can be transmitted along the food chain to human, which make the situation of AMR of *Salmonella* extremely severe. Therefore, the use of antibiotics should be further standardized and the AMR monitoring of *Salmonella* should be strengthened in the future.

The increasingly serious AMR of *Salmonella* has an adverse effect on the clinical treatment of salmonellosis. The biochemical AMR mechanisms of *Salmonella* are as follows: (1) Produce inactivating enzymes to destroy antibiotics; (2) Reduce the permeability of the bacterial outer membrane; (3) Strengthen the efflux of the active efflux pump to transport antibiotics out of the cell; (4) To modify the action target of antibiotics; (5) Target gene mutation. The serotypes or AMR genes can be retrieved and analyzed through the genome-based genotyping method and WGS data. The development of bioinformatics technology provides a new perspective and method for clarifying AMR mechanisms and detecting AMR.

To a certain degree, the AMR in *Salmonella* can be eliminated or reversed by traditional Chinese medicine or traditional Chinese medicine active ingredients. Some traditional Chinese medicines have good reversal effects on resistance of *Salmonella* isolates. By eliminating the resistant plasmids, Chinese herbal medicines can reduce

AMR of *Salmonella* strains and reduce the selection pressure of bacteria. Therefore, some traditional Chinese medicines, as an alternative to antimicrobial agents or a promoter of antimicrobial agents have important significance for the prevention and treatment of *Salmonella* infectious diseases.

Acknowledgements

The authors would like to acknowledge Wei Mao and Weiguang Zhou, Professors of Veterinary Medicine of Inner Mongolia Agriculture University. They provided a large amount of information on the molecular epidemiology of zoonotic pathogens. In addition, JinShan Cao, the senior vice mayor of Tongliao City in the Inner Mongolia Autonomous Region is acknowledged for proposing the new design ideas before writing the manuscript. During the process of manuscript's revision, some valuable suggestions were given by Professor Cao. The authors also gratefully acknowledges the support from Inner Mongolia Science and Technology major project (2021ZD0013).

Author details

Hongxia Zhao
College of Veterinary Medicine, Inner Mongolia Agriculture University, Huhhot, Inner Mongolia, Peoples' Republic of China

*Address all correspondence to: 18947199590@163.com

References

[1] Tindall BJ, Grimont PAD, Garrity GM, et al. Nomenclature and taxonomy of the genus *Salmonella*. International Journal of Systematic and Evolutionary Microbiology. 2005;**55**(Pt 1):521-524

[2] Gal-Mor O, Boyle EC, Grassl GA. Same species, different diseases: How and why typhoidal and non-typhoidal *Salmonella* enterica serovars differ. Frontiers in Microbiology. 2014;**4**(5):391

[3] Lamas A, Miranda JM, Regal P, et al. A comprehensive review of non-entericasubspecies of *Salmonella* enterica. Microbiological Research. 2018;**206**:60-73

[4] Giammanco GM, Pignato S, Mammina C, et al. Persistent endemicity of *Salmonella* bongori 48:z35:- in southern Italy: Molecular characterization of human, animal, and environmental isolates. Journal of Clinical Microbiology. 2002;**40**:3502-3505

[5] Shi C, Singh P, Ranieri ML, et al. Olecular methods for serovar determination of *Salmonella*. Critical Reviews in Microbiology. 2015;**41**:309-325

[6] Brenner FW, Villar RG, Angulo FJ, et al. *Salmonella* nomenclature. Journal of Clinical Microbiology. 2000;**38**:2465-2467

[7] Galanis E, Wong DMALF, Patrick ME, et al. Web-based surveillance and global *Salmonella* distribution, 2000-2002. Emerging Infectious Diseases. 2006;**12**(3):381-388

[8] Xu YM, Tao S, Hinkle N, et al. *Salmonella*, including antibioticresistant *Salmonella*, from flies captured from cattle farms in Georgia, USA. The Science of the Total Environment. 2018;**616**(617):90-96

[9] Kirk MD, Pires SM, Black RE, et al. World health organization estimates of the global and regional disease burden of 22 foodborne bacterial, protozoal, and viral diseases, 2010: A data synthesis. PLoS Medicine. 2015;**12**(12):e1001921

[10] Pan H, Paudyal N, Li XL, et al. Multiple food-animal-borne route in transmission of antibiotic-resistant *Salmonella* Newport to humans. Frontiers in Microbiology. 2018;**9**:23

[11] Marder EP, Cieslak PR, Cronquist AB, et al. Incidence and trends of infections with pathogens transmitted commonly through food and the effect of increasing use of culture-independent diagnostic tests on surveillance: Foodborne diseases active surveillance network, 10 US Sites, 2013-2016. Morbidity and Mortality Weekly Report. 2017;**66**(15):397

[12] Centers for Disease Control and Prevention. Surveillance for Foodborne Disease Outbreaks United States, 2015: Annual Report[R/OL]. Atlanta: CDC; 2017

[13] Liu L, Zhou PH, Gao H, et al. Research Progress on contamination and ldentification of *Salmonella* in food. Modern Food. 2023;**29**(13):107-109

[14] Haiyan X, Weibing Z, Kai Z, et al. Characterization of *Salmonella* serotypes prevalent in asymptomatic people and patients. BMC Infectious Diseases. 2021;**21**(1):632

[15] Rafaela GF, Denes KA, Adelino CN, et al. Worldwide epidemiology of *Salmonella* Serovars in animal-based

foods: A Meta-analysis. Applied and Environmental Microbiology. 2019;**85**(14):e00591-e00519

[16] Shah DH, Paul NC, Sischo WC, et al. Population dynamics and antimicrobial resistance of the most prevalent poultry-associated *Salmonella* serotypes. Poultry Science. 2017;**96**(3):687-702

[17] Ghoddusi A, Nayeri Fasaei B, Zahraei Salehi T, et al. Serotype distribution and antimicrobial resistance of *Salmonella* isolates in human, chicken, and cattle in Iran. Archives of Razi Institute. 2019;**74**(3):259-266

[18] Zhang CP, Song L, Cui MQ, et al. Serotype distribution and antimicrobial susceptibility to colistin of *Salmonella* spp. from chicken in China. Chinese Journal of Veterinary Drug. 2018;**52**(1):13-18

[19] Martins RM, Zimbardi ASD, Munive VKN, et al. Overview of antimicrobial resistance and virulence factors in *Salmonella* spp. isolated in the last two decades from chicken in Brazil. Food Research International. 2022;**162**(PA):111955

[20] Saksena R, Nayyar C, Manchanda V. Six-year susceptibility trends and effect of revised Clinical Laboratory Standards Institute breakpoints on ciprofloxacin susceptibility reporting in typhoidal *Salmonella* in a tertiary care paediatric hospital in Northern India. Indian Journal of Medical Microbiology. 2016;**34**(4):520-525

[21] Cosby DE, Cox NA, Harrison MA, et al. *Salmonella* and antimicrobial resistance in broilers: A review. Journal of Applied Poultry Research. 2015;**24**(3):408-426

[22] Shu KE, Priyia P, Nurul SAM, et al. *Salmonella*:A review on pathogenesis, epidemiology and AMR. Frontier in Life Science. 2015;**8**(3):284-293

[23] Khan MI, Ochiai RL, Von Seidlein L, et al. Non-typhoidal *Salmonella* rates in febrile children at sites in five Asian countries. Tropical Medicine & International Health. 2010;**15**(8):960-963

[24] Cui HX, Zhang XL, Liao XG, et al. Analysis of drug resistance of food - borne Staphylococcus aureus and *Salmonella* in Henan province in 2010. Modern Preventive Medicine. 2013;**40**(02):320-323

[25] Michael GB, Schwarz S. Antimicrobial resistance in zoonotic nontyphoidal *Salmonella*: An alarming trend? Clinical Microbiology and Infection. 2016;**22**(12):968-974

[26] Dong MM, Wang XF, Xu YZ, et al. Isolation, identification and analysis of drug resistance of *Salmonella* pullorum. China Animal Husbandry & Veterinary Medicine. 2018;**45**(9):2637-2644

[27] Gupta SK, Sharma P, McMillan EA, et al. Genomic comparison of diverse *Salmonella* serovars isolated from swine. PLoS One. 2019;**14**(11):e0224518

[28] Zhong SH, Feng SW, Li J, et al. Investigation on serotype, drug resistance and drug resisitance gene of *Salmonella* in livestock and poultry products of Guangxi. China Animal Husbandry & Veterinary Medicine. 2018;**45**(3):770-780

[29] Su L, Chiu C, Chu C, et al. Antimicrobial resistance in nontyphoid *Salmonella* serotypes: A global challenge. Clinical Infectious Diseases. 2004;**39**(4):546-551

[30] Lee H, Su L, Tsai M, et al. High rate of reduced susceptibility to ciprofloxacin and ceftriaxone among

nontyphoid *Salmonella* clinical isolates in Asia. Antimicrobial Agents and Chemotherapy. 2009;**53**(6):2696-2699

[31] Hasan R, Zafar A, Abbas Z, et al. AMR among *Salmonella* enterica serovars Typhi and Paratyphi A in Pakistan (2001-2006). Journal of Infectious Developmental Ctries. 2008;**2**(04):289-294

[32] Jia HY, Gao LD, Guo YC, et al. Epidemiological characteristics of *Salmonella* in food of animal origin in Hunan province. Chinese Journal of Preventive Medicine. 2014;**48**(08):699-704

[33] Qin S, Shen Y, Ma K, et al. Analysis of drug resistance surveillance of foodborne pathogens in Jiangsu province in 2012. Jiangsu Preventive Medicine. 2014;**25**(01):28-30

[34] Zhou Y, Zhang XQ, Hou SP, et al. Serotype distribution and drug resistance of 76 strains of *Salmonella* in Guangzhou. Modern Preventive Medicine. 2014;**41**(02):349-350

[35] Zhang J, Zhang XQ, Hou SP, et al. Serotype distribution and drug resistance of *Salmonella* isolated from 566 diarrhea cases in Guangzhou. Practical Preventive Medicine. 2016;**23**(05):616-617

[36] Samuel H, Albert R, Peter D, et al. Serotypes and antimicrobial resistance in *Salmonella* enterica recovered from clinical samples from cattle and swine in Minnesota, 2006 to 2015. PloS One. 2016;**11**(12):e0168016

[37] Ding YY. Latest research progress of bacterial resistance mechanism at home and abroad. Modern Preventive Medicine. 2013;**40**(06):1109-1111

[38] Balleste DC, Sole M, Domenech O, et al. Molecular study of quinolone resistance mechanisms and clonal relationship of *Salmonella* enterica clinical isolates. International Journal of Antimicrobial Agents. 2014;**43**(2):121-125

[39] Zhan Z, Xu X, Gu Z, et al. Molecular epidemiology and antimicrobial resistance of invasive non-typhoidal *Salmonella* in China, 2007-2016. Infectious Drug Resistance. 2019;**12**:2885-2897

[40] Zhou J, Li N. Bacterial resistance mechanism and control countermeasures. Grassroots Medical Forum. 2006;**21**:1052-1053

[41] Khurshid N, Khan BA, Bukhair SW, et al. Extensively drug-resistant *Salmonella* typhi meningitis in a 16-year-old male. Cureus. 2019;**11**(10):e5961

[42] Han ZX. Pharmacodynamic Study of Colistin on Three Multidrug-Resistant Gram-Negative bacilli. Harbin: Northeast Agricultural University; 2012

[43] Wang QX, Zhao G, Wu HT, et al. Research progress of tetracycline. Modern Animal Husbandry Veterinarian. 2012;**10**:67-68

[44] Zhang B. Analysis of tetracycline resistance of avian *Salmonella*. Chinese Livestock and Poultry Breeding Industry. 2018;**14**(05):38

[45] Di WT, Du XW, Wu J, et al. Isolation and drug resistance analysis of *Salmonella* from swine. Jiangsu Agricultural Science. 2014;**42**(10):278-280

[46] Li F, Wei SS, Liu JH, et al. Detection of tetracycline resistance and resistance genes in 247 strains of swine *Salmonella*. Chinese Journal of Veterinary Medicine. 2019;**39**(10):1927-1933

[47] Si XF, Lu GL. Hazards and countermeasures of veterinary quinolones residues. Animal Qauarantine in China. 2008;**25**(9):16-17

[48] Yao SX, Hao RE, Wang Y, et al. Study on molecular typing and drug resistance of *Salmonella* in Shanxi Province from 2014 to 2017. Chinese Journal of Zoonoses. 2021;**37**(9):815-820

[49] Zhang ZF, Yang JX, Li CC, et al. Analysis of Ciprofloxacin Resistance in *Salmonella* enteritidis. Wuhan: Chinese Society of Food Science and Technology; 2019. p. 11

[50] Li SM, Hao HL, Ning HM, et al. SSCP analysis of fluoroquinolone resistance gene mutation in swine *Salmonella*. Guangdong Agricultural Science. 2012;**20**:139-141

[51] Wu CQ, Feng J, Sen Y, et al. Research progress on classification and detection methods of antibiotic residues in animal-derived foods. Journal of Food Safety and Quality Inspection. 2019;**10**(21):7126-7132

[52] Wang L, Lin H, Cao LM. Detection methods and research progress of aminoglycoside antibiotics in animal derived food. Southern Fisheries. 2006;**1**:76-79

[53] Zhang GX. Preparation and Preliminary Application of Streptomycin Monoclonal Antibody. Chongqing: Southwest University; 2006

[54] Guan KK, Tan M, Wang L, et al. Antimicrobial susceptibility test for aminoglycosides, quinolones and β-lactams antibiotics against *Salmonella* isolated from diseased Yaks and Tibetan pigs. Chinese Journal of Animal Infectious Diseases. 2023;**31**(02):176-180

[55] Zhang CC, Zhao YL, Zhou Q, et al. Isolation, identification and drug resistance analysis of pathogenic *Salmonella* from chicken. Chinese Journal of Veterinary Drugs. 2012;**46**(5):11-13

[56] Huang K, Chen SJ, Huang J, et al. Drug resistance analysis of *Salmonella* from animal origin and identification of florfenicol resistance genes. Chinese Animal Husbandry Veterinarian. 2015;**42**(2):459-466

[57] Mondal T, Khan MSR, Alam M, et al. Isolation: Identification and characterization of *Salmonella* from duck. Bangladesh Journal of Veterinary Medicine. 2008;**6**(1):7-12

[58] Li F, Luo XW, Liu JH, et al. Detection and analysis of multidrug efflux pump oqxAB and florfenicol resistance gene floR in swine *Salmonella*. Journal of Jiangxi Agriculture. 2018;**30**(11):82-85

[59] Zhao HX, Wang XK, Song C, et al. Isolation, identification and drug resistance test of pathogenic bacteria of cow endometritis in Inner Mongolia. Chinese Journal of Veterinary Medicine. 2022;**58**(05):58-63

[60] Zhao X, Gao JP, Zhang L, et al. Isolation, identification and drug sensitivity test of pathogenic bacteria of cow endometritis in Kuitun area of Xinjiang. Contemporary Animal Husbandry. 2018;**36**:27-29

[61] Lauderdale T, Aarestrup FM, Chen P, et al. Multidrug resistance among different serotypes of clinical *Salmonella* isolates in Taiwan. Diagnostic Microbiology and Infectious Disease. 2006;**55**(2):149-155

[62] Parry CM, Threlfall EJ. Antimicrobial resistance in typhoidal and nontyphoidal salmonellae. Current Opinion in Infectious Diseases. 2008;**21**(5):531-538

[63] Ceyssens PJ, Mattheus W, Vanhoof R, et al. Trends in serotype distribution and antimicrobial susceptibility in *Salmonella* enterica isolates from humans in Belgium, 2009 to 2013. Antimicrobial Agents and Chemotherapy. 2015;**59**(1):544-552

[64] Bai L, Zhao J, Gan X, et al. Emergence and diversity of *Salmonella* enterica Serovar Indiana isolates with concurrent resistance to ciprofloxacin and cefotaxime from patients and food-producing animals in China. Antimicrobial Agents and Chemotherapy. 2016;**60**(6):3365-3371

[65] Bai L, Lan R, Zhang X, et al. Prevalence of *Salmonella* isolates from chicken and pig slaughterhouses and emergence of ciprofloxacin and cefotaxime co-resistant *S. enterica* serovar Indiana in Henan, China. PLo S One. 2015;**10**(12):e0144532

[66] Le HS, Harrois D, Bouchrif B, et al. Highly drug-resistant *Salmonella* enterica serotype Kentucky ST198-X1: A microbiological study. The Lancet Infectious Diseases. 2013;**13**(8):672-679

[67] Jiang HX, Song L, Liu J, et al. Multiple transmissible genes encoding fluoroquinolone and third-generation cephalosporin resistance co-located in non-typhoidal *Salmonella* isolated from food-producing animals in China. International Journal of Antimicrobial Agents. 2014;**43**(3):242-247

[68] Mc DPF, Walker RD, White DG. Antimicrobials: Modes of action and mechanisms of resistance. International Journal of Toxicology. 2003;**22**(2):135-143

[69] Boerlin P, Reid-Smith RJ. Antimicrobial resistance: Its emergence and transmission. Animal Health Research Reviews. 2008;**9**(2):115-126

[70] Foley SL, Lynne AM. Food animal-associated challenges: Pathogenicity and antimicrobial resistance. Journal of Animal Science. 2008;**86**(14 suppl):E173-E187

[71] Frye JG, Jackson CR. Genetic mechanisms of antimicrobial resistance identified in *Salmonella* enterica, Escherichia coli, and Enteroccocus spp. isolated from U.S. food animals. Frontiers in Microbiology. 2013;**4**:135

[72] Carattoli A. Resistance plasmid families in Enterobacteriaceae. Antimicrobial Agents and Chemotherapy. 2009;**53**(6):2227-2238

[73] Glenn LM, Lindsey RL, Folster JP, et al. Antimicrobial resistance genes in multidrug-resistant *Salmonella* enterica isolated from animals, retail meats, and humans in the United States and Canada. Microbial Drug Resistance. 2013;**19**(3):175-184

[74] Skov R, Matuschek E, Sjolund KM, et al. Development of a pefloxacin disk diffusion method for detection of fluoroquinolone-resistant *Salmonella* enterica. Journal of Clinical Microbiology. 2015;**53**(11):3411-3417

[75] Falagas ME, Karageorgopoulos DE. Extended-spectrum β-lactamase-producing organisms. Journal of Hospital Infection. 2009;**73**(4):345-354

[76] Deng Y, Bao X, Ji L, et al. Resistance integrons: Class 1, 2 and 3 integrons. Annals of Clinical Microbiology and Antimicrobials. 2015;**14**:45

[77] Vo AT, Duijkeren E, Gaastra W, et al. Antimicrobial resistance, class 1 integrons, and genomic island 1 in *Salmonella* isolates from Vietnam. PLoS One. 2010;**5**(2):e9440

[78] Li J, Zhang XH, Fu C. Advances in the study of the antibacterial effect of traditional Chinese medicine. China pharmaceutical. 2014;**23**(2):90-93

[79] Zuo GY, Wang GC, Zhao YB, et al. Screening of Chinese medicinal plants for inhibition against clinical isolates of methicillin-resistant Staphylococcus aureus (MRSA). Journal of Ethnopharmacology. 2008;**120**(2):287-290

[80] Liu BX, Zhao AQ, Liu CC, et al. Antibiotic susceptibility test of *Salmonella* diarrhea in calves and in vitro Bacteriostasis test of traditional Chinese medicine. Chinese Journal of Veterinary Medicine. 2021;**57**(3):26-30

[81] Li D, Yu D, Zeng FL, et al. Bacteriostatic effect of five kinds of traditional Chinese medicine on *Salmonella* cholerae. Chinese Journal of Veterinary Medicine. 2019;**39**(2): 318 322

[82] Ma C. Elimination of Three Bacterial Resistant Plasmids by Different Chinese Herbs. Chengdu: Sichuan Agricultural University; 2011

[83] Cao ZH. Effect of Traditional Chinese Medicine on β - Lactam Drug Resistance of *Salmonella*. Guiyang: Guizhou University; 2017

[84] Clinical and Laboratory Standards Institute. Performance standards for anfimicrobial suscepfibity testing. 28th ed. CLSI supplement M100. Wayne, PA: Clinical and Laboratory Standards Institute; 2018

[85] Zankari E, Hasman H, Cosentino S, et al. Identification of acquired antimicrobial resistance genes. The Journal of Antimicrobial Chemotherapy. 2012;**67**(11):2640-2644

[86] Tagini F, Greub G. Bacterial genome sequencing in clinical microbiology: A pathogen-oriented review. European Journal of Clinical Microbiology & Infectious Diseases. 2017;**36**(11):2007-2020

[87] Gordon NC, Price JR, Cole K, et al. Prediction of Staphylococcus aureus antimicrobial resistance by whole-genome sequencing. Journal of Clinical Microbiology. 2014;**52**(4):1182-1191

[88] Stoesser N, Batty EM, Eyre DW, et al. Predicting antimicrobial susceptibilities for Escherichia coli and Klebsiella pneumoniae isolates using whole genomic sequence data. The Journal of Antimicrobial Chemotherapy. 2013;**68**(10):2234-2244

[89] Zhang SK, Bakker HC, Li ST, et al. SeqSero2: Rapid and improved *Salmonella* serotype determination using whole-genome sequencing data. Applied and Environmental Microbiology. 2019;**85**(23):e01746-e01719

[90] Xu F, Ge CT, Luo H, et al. Evaluation of real-time nanopore sequencing for *Salmonella* serotype prediction. Food Microbiology. 2020;**89**:103452

[91] Zhang SK, Yin YL, Jones MB, et al. *Salmonella* serotype determination utilizing high-throughput genome sequencing data. Journal of Clinical Microbiology. 2015;**53**(5):1685-1692

[92] Diep B, Barretto C, Portmann AC, et al. *Salmonella* serotyping; comparison of the traditional method to a microarray-based method and an in silico platform using whole genome sequencing data. Frontiers in Microbiology. 2019;**10**:2554

[93] Robertson J, Yoshida C, Kruczkiewicz P, et al. Comprehensive assessment of the quality of *Salmonella*

whole genome sequence data available in public sequence databases using the *Salmonella* in silico typing resource (SISTR). Microbial Genomics. 2018;**4**(2):e000151

[94] Banerji S, Simon S, Tille A, et al. Genome-based *Salmonella* serotyping as the new gold standard. Scientific Reports. 2020;**10**:4333

[95] Yachison CA, Yoshida C, Robertson J, et al. The validation and implications of using whole genome sequencing as a replacement for traditional serotyping for a national *Salmonella* reference laboratory. Frontiers in Microbiology. 2017;**8**:1044

[96] Khezri A, Avershina E, Ahmad R. Plasmid identification and plasmid-mediated antimicrobial gene detection in Norwegian isolates. Microorganisms. 2020;**9**(1):52

[97] Neuert S, Nair S, Day MR, et al. Prediction of phenotypic antimicrobial resistance profiles from whole genome sequences of non-typhoidal *Salmonella* enterica. Frontiers in Microbiology. 2018;**9**:592

[98] Zankari E, Hasman H, Kaas RS, et al. Genotyping using whole-genome sequencing is a realistic alternative to surveillance based on phenotypic antimicrobial susceptibility testing. The Journal of Antimicrobial Chemotherapy. 2013;**68**(4):771-777

[99] Zhu YT, Lai HM, Zou LK, et al. Antimicrobial resistance and resistance genes in *Salmonella* strains isolated from broiler chickens along the slaughtering process in China. International Journal of Food Microbiology. 2017;**259**:43-51

[100] Walker TM, Kohl TA, Omar SV, et al. Whole-genome sequencing for prediction of *Mycobacterium tuberculosis* drug susceptibility and resistance: A retrospective cohort study. The Lancet Infectious Diseases. 2015;**15**(10):1193-1202

[101] Li XR, Zhang RH, Yang Y, et al. Whole genome sequencing of foodborne pathogens and global data sharing development. Chinese Journal of Food Hygiene. 2020;**32**(3):339-344

[102] Li X, Qiang B, Xu ZZ, et al. Application of whole genome sequencing in bacterial antimicrobial resistance. Chinese Journal of Zoonoses. 2016;**32**(8):696-699

[103] Zhao S, Tyson GH, Chen Y, et al. Whole-genome sequencing analysis accurately predicts antimicrobial resistance phenotypes in Campylobacter spp. Applied and Environmental Microbiology. 2016;**82**(2):459-466

[104] McDermott PF, Tyson GH, Kabera C, et al. Whole-genome sequencing for detecting antimicrobial resistance in nontyphoidal *Salmonella*. Antimicrobial Agents and Chemotherapy. 2016;**60**(9):5515-5520